LAND OF MILK AND HONEY

LAND OF MILK AND HONEY

DIGRESSIONS OF A RURAL DISSIDENT

JAMIE BLACKETT

Quiller

Also by Jamie Blackett:

The Enigma of Kidson: The Portrait of an Eton Schoolmaster (2017)
Red Rag to a Bull: Rural Life in an Urban Age (2018)
On Wilding (2022)

First published in the UK in 2022
by Quiller, an imprint of Amberley Publishing Ltd

British Library Cataloguing-in-Publication Data
A catalogue record for this book is available from the British Library

ISBN 978 1 84689 366 7 (hardback)
ISBN 978 1 84689 367 4 (ebook)

British Library Cataloguing in Publication Data.
A catalogue record for this book is available from the British Library.

1 2 3 4 5 6 7 8 9 10

Typesetting by SJmagic DESIGN SERVICES, India.
Printed in the UK.

Quiller

An imprint of Amberley Publishing Ltd
The Hill, Merrywalks, Stroud, GL5 4EP
Tel: 01453 847800
E-mail: info@quillerbooks.com
Website: www.quillerpublishing.com

In memory of my father,
Beachie Blackett,
1939–2019

Contents

Prologue

June 2019 was the lunar equivalent of an annus horribilis. The price of beef plumbed new depths just as the bulls came fat, driven down further no doubt by the vegan popular front. The holiday cottage booking system caused my blood pressure to reach new heights, exacerbated by algorithms created by Californian dotcoms; but worse, much worse, my old dad died. It was his pipe that got him in the end; that and the fags, which caused his old lungs to give up the unequal struggle, a medical fact that I tried to impress on his eleven grandchildren. I suppose you could say that it was an occupational hazard, given that for years the pipe had warded off the midges on evenings spent fishing on the River Borgie, and latterly the Nith, or waiting for a duck in the gloaming of a September evening after harvest, and it kept him warm in the punt as he edged it past icebergs in January dawns stalking geese on the Solway. Though he had given it up ten years ago on doctor's orders, I found it in a drawer and stuck it in his coffin with his old leather pouch of whisky flake. It still had the empty brass case of a .303 round he had picked up off a firing range in Kenya in 1960 and used to tamp the baccy into the bowl of his pipe for fifty years thereafter. His declaration that he had 'no live rounds or empty cases in his possession' that day had been a fib. I like to think he is up there now, wreathed in blue smoke, puffing away contentedly, rod in hand.

His last conversation with me was about hunting. And his pipe accompanied him into the hunting field as well. If there is still hunting in the Elysian Fields, and I am sure there is, he will be there on his good mare Twiggy, pipe jutting out beneath his top hat at each check or smouldering in his coat pocket when the action starts. As he lay in hospital in Oxford, the life slowly fading out of him, he gazed out of the window across the city to the patchwork of green fields beyond. 'I like to imagine a good run across those fields, it looks good hunting country.'

We brought him home to Galloway to be buried. Fittingly it was a busy farming day, after a catchy spell of weather there was a window to get on and cut silage. I hesitated before allowing the contractors to come on the day of the funeral, then thought Dad would have insisted that we cracked on, so made a plan for them to go for it with the aim of sheeting the pit the day after, it would be good for us to be kept busy. So, as we sadly made our way towards the church in Dumfries we passed frenzied activity: the chopper munching its way through thick rows of mown grass in the field next to the kennels, and tractors thundering round the roads leading silage to the pit. It was a grassy year and our first cut was the heaviest ever. Dad had seen it on his last day with us ten days before and had glowed with pleasure as he gave me rare praise for the crop. It all served to emphasise that life goes on.

When our friend Maggie Gladstone played the 'Eton Boating Song' on the organ while the congregation took their seats, there was for once a genuine 'hay harvest breeze'. And the metaphor being 'gathered' seemed very appropriate as his grandsons carried him in feet first. Everything had gone according to plan. He would have been quietly pleased by 'a good turnout' and chuckled at his friend Andrew Duncan's eulogy recalling their many days shooting together, and I hope his chest swelled inside the coffin as he was carried out to the strains of the regimental march, 'Milanollo', by a different team: Davie, Graham, Ainslie, Peter, John and Raymond, all of whom had worked on the estate at one time or another.[1] There was a feeling of relief as we loaded him onto the hearse and headed out of the town for his last journey back to 'God's Acre', the little family cemetery at Arbigland, to be buried alongside previous

---

1. Dad had made it clear that he did not want to be wheeled out like a cake on a trolley.

generations and, as importantly, his dogs: spaniels and a long line of huge yellow Labradors already buried around the walls.

It was only at the lychgate that it all went pear-shaped. If you had to nominate the five most dispiriting words in the English language, 'the grave has fallen in' take some beating. But then Dad's life never went entirely according to plan. I could almost hear him roaring with laughter and declaring that it was, 'like something out of *The Irish RM*' as I hastily convened an 'O Group'[2] with the vicar, agreed a contingency plan, and we carried on with the 'dust to dust' bit with his coffin placed on the grass, and listened to Andrew our huntsman blowing 'Gone Away' to speed him on his way before turfing Dad over and heading for the wake.

* * *

Victorian novelists loved to write about the death of the squire – or laird in this case – with all the drama of the reading of the will and the seismic effects on the family and the community on the estate. That is rarely the case these days as the effects of inheritance tax planning and longevity mean that most landowners try to hand over long before they die, as my father had done to me nearly twenty years before. He had then carried on living here until my stepmother's illness forced a move to be closer to my sisters, fulfilling the sort of laird emeritus role touchingly played by Richard Briers in the television series *Monarch of the Glen*. The same role that I remember my own grandfather filling. Driving around the estate, tweed cap pulled down over his eyes, fag in mouth, commenting acidly on any lapses in standards; chairing the Community Council in autocratic *Vicar of Dibley* style and ensuring that the village's poppy collection exceeded the previous year's record.

It should not, therefore, have felt like a moment of epochal change. I had been running things for two decades since leaving

---

2. O Group stands for Orders Group, without which nothing in the army happens. Dad continued to run his life that way as a civilian and I was always being 'blown for' for an O Group.

the army – a story told in my previous book *Red Rag to a Bull*[3] –
and will, God willing, have another one to go before it is my son
Oliver's turn to take on the responsibilities. But it felt as if the
ground was being wrenched from under my feet and I suffered from
a deep depression that went well beyond bereavement.

For twenty years I had gradually built a business that could
sustain us, doing up properties, starting a successful holiday
cottage business, breeding a pedigree beef herd and initiating the
resurrection of the Dumfriesshire and Stewartry Foxhounds on
the estate. We had started to make good profits and I was at last
earning more than I had in the army. At the end of *Red Rag to a
Bull*, with egregious conceit I had finished with a poem that ended
with the words, 'All will be well.' But almost as soon as I had put
my pen down nemesis began to overcome my hubris. The groove
I thought I had carved out started to feel like a rut. Cheap flights
to the sun were making holiday cottages a hard sell, the price of
beef was collapsing as our costs kept going up. I had hidden it
from Dad, as I didn't want to worry him, but we were starting to
lose money and every waking moment was spent worrying about
cashflow as the potholes in the drive deepened and the fabric of the
estate started to unravel before my eyes for lack of funds. I felt like
Mr Micawber in *David Copperfield*:

> Annual income twenty pounds, annual expenditure nineteen
> nineteen six, result happiness. Annual income twenty pounds,
> annual expenditure twenty pounds ought and six, result misery.

But I knew that hoping that 'something will turn up', like
Mr Micawber, was not an option and would result in disaster as we
would rapidly be unable to keep up the various mortgage payments.
Brexit was coming down the tracks at us like an express train and
I was chained to the tracks. If we were not making money then,
how would we survive when we left the EU, our subsidies were
removed and cheap imports flooded the UK to fulfil the Brexiteers'
promise of cheaper food?

---

3. Available on Amazon and in all good bookshops, a steal at £20 if you want to
   start at the beginning.

Beneath it all Scotland was feeling like an increasingly unhappy place. The decisive vote in 2014 to remain part of the United Kingdom was being ignored by the SNP on the pretext that Brexit changed everything. The 'once in a generation' promise was being rescinded as they pressurised Westminster for another vote. Every time Nicola Sturgeon opened her mouth it felt as though we were losing customers. Traditionally 80 per cent of our holiday cottage guests came from England and only 20 per cent from Scotland. Some weeks those proportions were being reversed. And the Scotch beef premium, which had given Scottish farmers as much as 15 per cent more for their meat than their English competitors, was fast disappearing.[4] It was dawning on me that our farming and tourism brands that had been steadily built up by Scottish farming families were now being trashed in the eyes of English housewives by the steady drip of nationalist Anglophobia in the media. Communities were becoming divided once again and the poison of separatism seemed to infect every political decision. I had served in Northern Ireland and seen how the local people kept their heads down, intimidated by nationalists. I could feel Scotland going the same way.

I was also depressed at the thought of losing the hunt. After the law in Scotland changed in 2001, to make it so that foxes had to be flushed to guns by hounds, we had started a new hunt in 2005 and had managed to keep going under the new rules. It had created a tight-knit community with friendships forged across three counties[5] and every walk of life. Our lives had revolved around not just the hunting but hunt balls, point-to-points and raft races – all the fundraising events that raise money for the hunt, but, more importantly, maintain morale in rural communities. Now Nicola Sturgeon was threatening an outright ban on hunting.[6]

It was as if everything in my life was falling apart and all my efforts of the last twenty years were coming to nought. I have

---

4. At the time of writing it has been reversed completely.

5. Dumfriesshire, the Stewartry of Kirkcudbright and Wigtownshire, which together form Dumfries and Galloway.

6. The current proposal would make it impossible by insisting on only two hounds and also outlaw trail hunting.

always compared aspects of farming – the feeling of two steps forward two back – to the story of Sisyphus, the mythical Greek condemned to push a boulder up a hill forever. It now felt as though the boulder might overcome me and crash to the bottom of the hill, taking me with it. The stress caused a large black dog to break loose inside my head and it chased me night and day so that sleep was impossible and even violent exercise could not shift the adrenaline in my system. After weeks without sleep I started to inhabit nightmare worst-case scenarios and became obsessed with the thought of having to sell up and move away, of being cursed with being the one who dropped the baton and severed our connection with the land after at least seven centuries of land ownership in my family. By cruel coincidence that long hot summer of 2019 made Arbigland look bewitchingly beautiful, the view across the Solway to the mountains in the Lake District took on deeper shades of blue in the midsummer sun and the roses against the mellow stonework of our home, the House on the Shore, bloomed as never before. The thought of falling from paradise haunted me.

I had to find a way through it all. But how?

CHAPTER 1

Rock Bottom

January 2019

'I'm sorry I can't give you a date. They are full this week and next. Maybe the week after but I can't promise anything.'

'Okay Guy, well please do your best, we really need them away.'

There is a large and hyperactive butterfly fluttering around in my stomach as I put my mobile back in my pocket after talking to the Farmstock buying agent. Davie and I exchange shrugs and I can see that he is as worried as I am. Davie is my stockman, mechanic, solver of problems, fixer of breakages and right-hand man. We are standing by the cattle sheds watching the fat cattle eating. Every mouthful is pushing us further into debt. The bulls are in peak condition. The longer we keep them the more it costs, and they will put on fat now so that when their carcasses are graded at the abattoir we will receive less pence per kilo. There is a short window to have them at the right specification before they go over the sixteen months age limit and over-weight and are worth much less.

I look around at the cattle sheds we have built, the concrete we have laid, the feed barriers Davie has welded with great care, and most of all at the cattle. After fourteen years of hard graft there are 200 cows now and 400 youngstock, all of which we have bred ourselves as the original forty I bought have all gone now, which means that Davie has pulled a fair few of them into the world himself and nurtured them as calves. Suddenly for the first time I am wondering if it has all been worth it. If no one wants our beef and we can't even book them into an abattoir to be killed.

Everyone is in the same boat and our farming neighbours are equally worried. 'Up corn, down horn' is an old farming expression that often accurately reflects the agricultural business cycle. It also means that we are hit with a double whammy: low output prices and high input costs – of grain for fattening the cattle and fertiliser for making the grass grow, which usually follows corn up in price. The problem is that, as the supermarkets have fussed about the imminence of Brexit and the possibility of a 'no-deal' exit, the supply chain has filled every chiller with beef, most of it brought in from Ireland and Poland where the big three Irish meat processors who dominate the UK beef market also have abattoirs and cutting plants. And now they don't need Scotch beef until further notice. This has never happened before. There have been times of poor demand when there has been a backlog and a queue to book cattle in for killing but it has never been like this. And usually it has been driven by the value of sterling being too high, this is happening when the pound is near an all-time low. Suddenly the reassurances that doing without EU subsidies would be more than offset by devaluation don't look so reassuring. It takes between twenty-five and thirty-nine months between turning the bulls out with the cows and selling the resulting progeny to the abattoirs. When some of these cattle were conceived we hadn't even voted to leave the EU.

The final twist of the knife is 'Veganuary', which started with a CBE for a certain high-profile vegan BBC presenter whose main mission seems to be to drive livestock farmers out of business. The Campaign for Random Accusations against Pastoralists – I'll spare you the acronym – has intensified this year. And worryingly it has changed the buying habits of yet more consumers with some reports, possibly fake news, stating exponential year on year growth of non-meat products, including, bizarrely, petfood. It reminds me of Kipling's poem 'If', especially the bit about 'being lied about, don't deal in lies/Or being hated, don't give way to hating'. For there is no doubting the lies and the visceral hatred that some vegans have for British farming (as opposed to nice friendly soya cultivation in the Amazon basin). They are impervious to reason: I have just made the mistake of getting into a Twitter spat with a vegan blogger who refused to accept that leather is a by-product of the livestock industry. And they won't let the facts get in the way as they ruthlessly bend the scientific community and the media to their will. If you want a masterclass in black propaganda, watch *Cowspiracy* on Netflix. It bewails the volume of water required to produce a kilo of beef. Really? Have these people ever been to Galloway? The trouble is if you are facing a hosepipe ban in Kent it is all too believable.

February 2019

The spine-chilling sound of a February night comes from foxes barking as they tryst on the bank above the house. There was a time when it would have had us rushing out with the rifle and a lamp. We used to have zero tolerance of foxes but we have been more forgiving since we have had the hunt kennelled here, and sought to maintain a balance. The hunt has accounted for several here this winter and I have not begrudged the occasional pheasant to the surviving vulpine population. Not so elsewhere. Some parishes are completely fox free. The extraordinary leap in the technology of night sights has tilted the odds firmly in the keeper's favour and there has been a marked decrease in the fox population over the last decade. I am not sure it is an entirely healthy development, except on grouse moors where it is essential to keep the fox population down to conserve all the ground-nesting birds, especially waders.[7]

February also means that pheasants have reappeared everywhere with cocks starting to spar among the snowdrops. Where were you when we needed you in that blank drive you old bastards? It always amazes me how they seem to know that hostilities have ceased. The geese seem to know as well. We only had barnacles here for the last few weeks of the season, flaunting their protected status as they munch what little grass remains for the sheep. Now some pinkfeet have returned. The freezer is crammed full of ten months' worth of 'pheasant surprise' culminating in the first shooting lunch of next season, if Sheri has not divorced me by then on the grounds of unreasonable demands on her culinary skills – and her good nature. Diehards who campaign to extend the shooting season to the end of February tend to be bachelors in my experience.

---

7. The British Mammal Society reports a sharp decline in the fox population in England and Wales since the hunting ban.

We have finally managed to get the remaining fat cattle away, but at a loss. I am racking my brains to try and find a solution to our deepening crisis. One half of my brain is telling me that this is a temporary blip, we need to stick with it and all will be well. The other is telling me that this is a wake-up call. Brexit is going to force huge changes and being an ostrich will jeopardise our future here.

March 2019

March. Here the fields that, God willing, will be hidden by dense crops in a matter of weeks, are snooker-table bare, not even striped by the roller yet, especially where the geese are taunting me by grazing through the day. Driving around, seeing the first sprays of blackthorn in the hedges and a reassuring crop of catkins for the birds this year, usually I will stop and search some of our bigger fields for hares dancing their March madness in the middle. There used to be at least half a dozen in each field when I was a child. But that was when the estate was intensively keepered and we still had a few coveys of grey partridges knocking about in the early seventies. We still see hares most days but of course they are much rarer now. If you drew two graphs and had one showing the decline in the number of hares, and the other showing the increase in the number of badgers and other predators over the same period, I wouldn't mind betting that seen side by side they would form a symmetrical V. 'Modern farming practices', the usual scapegoat, have, if anything become more wildlife friendly here since the seventies. Although there are some who take three cuts of silage, we don't and we reckon there is usually time for leverets to be up and off before our first cut in mid-June and then again before the second in August.

The hares no longer gather so conspicuously either. Perhaps they have adapted their behaviour to become more covert as the skies have filled with raptors. We no longer shoot hares here and the 'no ground game' edict seems to be fairly universal on shoots. But when my brother and I were learning to shoot my father could always guarantee that if we went out with the old .22 rifle we would get a stalk and then a shot. And sure enough, the Easter holidays when I was ten, the first thing I ever shot was a hare and

I can still remember the thudding of my heart as I stalked it and the intense concentration as I pulled the trigger, then the pride with which I carried it back, my cheeks sticky where my father had 'blooded' me, and the taste of it jugged a few days later.

One thinks of the hare as a gentle, passive creature but they can be fierce in defence of their young. I once witnessed an adult hare knocking seven bells out of a rook. There must have been leverets nearby and the rook had obviously gone too close; the two tussled together for several minutes, with the hare kicking it hard. Eventually the rook threw in the towel and flew away unsteadily. The hare seemed none the worse for her scuffle. Sadly it was before the days of camera phones or it would surely have gone viral.

I am worried that there have been hare poachers. I keep getting reports of strange vehicles and lights at night. It is one of the baleful consequences of the hunting ban. We all said at the time that a ban on coursing would be terrible for the hare population and it is. It is fifteen years since the last Waterloo Cup, the climax of an annual programme of organised and highly regulated coursing events that followed the principles of natural selection, culled a few of the weaker hares and provided the incentive for hare conservation. Poachers rubbed shoulders with landed gentry in a bibulous festival of one-nation Toryism. Since then the gentry have departed like Chesterton's last sad squires – leaving the fields to violent gangs of thugs who kill many more hares indiscriminately and terrorise farmers.

* * *

'Oh my God. She looks as if she has mad cow disease.'

The cow, a good-looking Sim-Luing,[8] one of the best in the herd, is shaking uncontrollably and staggering across the yard. Davie is having a well-earned break and Graham and I have brought the cows out of the shed to re-do their bedding. These crises always seem to happen when he is off. This is all we need. We watch helplessly as she writhes in agony and lurches across to one of the

---

8. A Luing cow put to a Simmental bull produces a Sim-Luing, one of the best hybrid suckler cows. Our herd contained a number of them.

feed troughs and climbs in. I run across to try and stop her getting cast and then something remarkable happens. With her front feet on the trough and her hind feet on the concrete she stretches herself to her fullest extent and tilts her head back then there is a click and she instantly returns to normal, climbs out of the trough and walks back across the yard as if nothing has happened. She must have had a trapped nerve in her back and fixed it for herself. I think of my own troubles with a bad back. I have learnt to do exercises whenever a slipped disc threatens. But in my case it was a physiotherapist who taught me how to deal with it. She had worked it out for herself.

* * *

The joy of watching our cattle is tempered by a gnawing unease in the pit of my stomach. The continuing crisis in the beef market is concentrating minds. The bullish farmers for whom normally 'every egg is a double yoker' are some of the worst affected. We are all looking out for each other, greetings are suffixed with 'Are you okay?', a question loaded with the knowledge that one farmer a week is taking his life in the UK. Mental health is no longer a taboo subject and the farming charities, overrun with providing unprecedented support to people close to the edge, are encouraging us to talk. Wherever I go the weary resignation is the same. It is focused on 'they'. 'They don't care about farming, they just want one big forest.' 'They (the supermarkets and beef processors) have sucked all the profit out of the job.' 'They are happy to burn fossil fuels flying avocados around the world then blame our cattle for climate change.' The excellent Minette Batters[9] spoke for us all when she urged celebrities to be careful of the consequences of their actions in the wake of yet another vegan publicity stunt, and warned of the dangers of a feeling of worthlessness on family farms. When I re-tweet Minette's comments on Twitter an anonymous vegan activist replies to me, without a shred of irony, 'Livestock farming will be done within the next decade ... Lab grown meat is the future.' Giving a talk to a group of farmers in Lochaber I attempt

9. President of the English NFU.

to cheer them up by reminding them that 'lab food' might not catch on. We have had artificial insemination for a couple of generations but the vast majority of couples still prefer the orthodox method of procreation.

It is at times like this that we question why we do it. It boils down to a visceral attachment to a business that is also a home and to animals that are part of our extended family. The wicked accusation that farmers don't care about their stock was answered powerfully with images posted on social media of Faye Russell from Derbyshire swimming in a river to save her sheep during Storm Dennis. All that could be seen at one stage was her woolly hat.

April 2019

It is a blessing that the clocks have sprung forward. One old cock pheasant insists on cock-cocking and beating his chest below our bedroom window the moment there is a glimmer of light so we now wake one hour later, not that I am sleeping much worrying about how we are going to keep going. It is an iconic sound of British spring yet our Celtic ancestors must have had an awful shock when their Roman landlords produced these strange Asian fowl. The cocks are looking magnificent with chests like the summer coats of bay horses.

Our exit from the European Union hangs over the farming community. It is a surreal time as we drill spring crops and go about calving and lambing not knowing what the future holds for British farming. The hopes of an end to bureaucratic regulations that persuaded many farmers to vote Leave are currently outweighed by fears of the end of family farms like ours in the face of a double whammy of reduced subsidies and cheap imports. Conversations with farming neighbours are accompanied by frequent shrugs and sighs. It is a rather joyless spring this year as a result of the gnawing doubts. We have tightened our belts and given up any idea of a holiday.

It is a welcome morale boost then, when my neighbour Jeremy, the sporting agent, proposes a day on the loch. He has inside information from our friend Puffin the fish farmer, who stocks the loch, that a consignment of rainbow has just been deployed.

Jezza has had a lean time of it after the dreadful grouse season of 2018. 'Three men in a boat, that's what we need. Dickie is coming, are you on?' And so like Jerome K. Jerome's three stalwarts we wobble out of the boathouse into the lapping waters of the loch with Ines, Jezza's black lab playing the part of Montmorency with her front paws eagerly perched on the bow. Jezza takes the first shift on the oars and mutters disparaging Luddite thoughts at one of the other boats as it slides past, its electric outboard churning the water noiselessly. Dickie is a slipper farmer[10] now and has a keen appreciation of the good things in life. He rummages in his creel. 'Time for a drink? It's gone 10.30. I think of this as a good mid-morning beer.' He produces a bottle of Old Crafty Hen.

Out in the middle we are able to appreciate the aching expectancy of a Scottish spring. There is a tentative hint of warmth in the breeze, a pair of kestrels is chasing each other through the tops of the birch trees and striped silage fields stretch away in the distance, green with promise, looking much longer than usual. I scan the loch with my binos, a bunch of tufted duck rises behind the island and everywhere there are pairs of birds in the eager pangs of courtship: swans, coot, mallard and the hooting intruders, Canada geese. 'No swallows yet.'

The hours fly past as we each catch a couple of glistening trout, giggling like ten-year-olds as we scrabble in the boat for landing net and priest. Then a leisurely lunch on pork pie and mustard washed down with claret on the island with its ruined chapel and thicket of untamed trees, before boarding the boat, a trifle unsteadily, the world put to rights, for the pull back to the boathouse. Then, out of the corner of my eye, a half-forgotten speck flickers across the water. 'Was that a swallow ...? Sand martins!' Small, brown and perfectly formed like some of the flies we have been using, there is no mistaking them as they soar and dive in joyful rediscovery of the loch. They have kept faith with us and made the long trip back from the Sahel. And suddenly life doesn't seem so bad after all.

* * *

---

10. A farmer who has ceased having stock of his own and lets his land out.

I can tell something is wrong as soon as I walk into the kitchen. Rosie has failed her driving test, one of life's vicissitudes, as I start to point out – I failed mine several times. But it was the manner in which she failed that is upsetting. The examiner had intimidated her as she drove around the streets of Castle Douglas.

'Where are you from? You don't sound Scottish.'

'Oh, well I am. I live locally.'

'Ah, so the kind of Scot who doesn't vote for independence then.'

He had put her off by grabbing the steering wheel, accusing her of being too close to some parked car, something she vehemently denies, and she had ended the test as a bag of nerves.

It has depressing echoes of Oliver's test a few years ago. When Sheri collected him from the test centre, his examiner told Sheri with what sounded like satisfaction that he had failed.

'Oh no, what a shame. It's so hard for young people living where we live if they can't drive.'

'Oh, I ken where you live.'

They say that the slow descent into nationalism has not been so keenly felt because of boiling frog syndrome – the frog does not know the water is getting hotter at first so that it is too late when it does. It's when an incident happens that violates one's familiar bubble that one is shocked to realise the cauldron of hatred around it. Nationalism needs an 'other' to thrive. In Germany it was the Jews, in Scotland it is the English, or Scots like our children, who sound English, who are 'othered'. The faux romance of the independence movement is not sustained by a love of Scotland but by a hatred of the English.

We curse ourselves for not paying Rosie's driving instructor to sit in the back during the test, something that is possible apparently. And we wonder about making an official complaint, but there doesn't seem much point, not when the bigotry comes right from the top.

* * *

A book comes in for review: *Green and Prosperous Land, A Blueprint for Rescuing the British Countryside* by Dieter Helm.

One of my favourite sayings is Jonathan Swift's, 'Whoever could make two ears of corn or two blades of grass to grow upon a spot of ground where only one grew before, would deserve better of mankind, and do more essential service to his country than the whole race of politicians put together'. It is a precept that has driven generations of farmers ever since he wrote it in 1726, but maybe not for much longer. During the Queen's reign average household spending on food in the UK has fallen from 40 per cent to 10 per cent and, despite the recent proliferation of food banks, we live in an age of plenty. In fact the food banks, recipients of food that would otherwise be wasted, are themselves evidence of surpluses that would once have been unimaginable. But here is a book that assumes that 'the possibility of Britain being cut off from foreign food supplies … is absurd: it is just silly'.

Dieter Helm is an Oxford Economics don and has the ear of government. He brings 'the dismal science' to bear in setting out a comprehensive blueprint for how we might reshape environmental policy after Brexit. He has a highly jaundiced view of modern agriculture coloured by the destructive conversion of his grandfather's 350-acre farm in Essex into one large arable field in the 1960s, and we are all tarred with the same brush. But it is hard not to agree with his central thesis that we have wreaked havoc on our wildlife since the Second World War in pursuit of cheap food and we should now look upon our countryside in terms of its natural capital and recalibrate support mechanisms to reward good ecology with subsidy, and tax pollution. Enlightened farmers are used to seeing the returns from the land in qualitative as well as quantitative terms, and many do want to farm less intensively to enjoy richer birdlife and wild flowers as long as there is still a living to be made. Professor Helm is fairer than many environmental activists in acknowledging that 'seventy years of subsidies have not brought prosperity to the bulk of farmers'. However, he acknowledges the possibility of 'an awful violence to our farmers' as many go out of business as a result of the reforms he proposes. His rather utopian solution of insisting on carbon taxes and welfare standards at the border to avoid simply exporting our production overseas may not stand up to the demands of the free traders, or indeed a porous Irish border.

For all its good intentions, the second half of the book descends into a depressingly interventionist agenda of Stalinist-lite five-year plans and bureaucratic controls for the countryside backed up by Orwellian policing of 'compliance' by drones and satellites. I wonder whether the word 'prosperous' in the title will apply not to farmers under his proposed regime but rather to civil servants through the extension of the Big State in partnership with its greedy and arrogant nephew Green quango.

Is this what we are hanging on for after Brexit?

***

'It's not good news I'm afraid. Are you sitting down?'

The voice at the other end of the line is friendly but hesitant, with the tone of someone who has something unpleasant to impart.

She goes on, 'I'm sorry but we have looked at it again and we can't pay you for the green manures. We accept that you did nothing wrong but the fact that the claim was made incorrectly means that it is being counted as an infringement so the money won't be paid.'

My head swims and I feel a chemical shock of anxiety dissolving my insides.

'But … Then we can't carry on … we were relying on this money being paid.'

'I know, I'm really sorry.' I know she is genuine. She is a farmer's wife who works part time for the department.[11] 'I'm afraid it gets worse. Because of the size of the claim it will incur penalties on some of your other claims and we will need to claw some of the money back from those as well. We don't know how much but we will let you know as soon as possible. You can appeal of course but that may take some time.'

There is no point in arguing. This has been going on for two years and it has achieved nothing so I thank her and put the telephone down.

Quos Deus vult perdere, prius dementat – Those whom God wishes to destroy, he first sends mad.

---

11. The Scottish Government Rural Payments and Inspections Division.

It started when I thought that I would be clever and plan our arable rotation around green manures paid for through subsidy. It was in effect a risk-free income; the payment equated to a bumper crop of wheat, it gave us a break from growing cereals and the soil was left in a much better condition afterwards as the deep-rooted legumes broke up the soil and fixed nitrogen. When they were ploughed in (hence green manure) they would enrich the soil with organic matter. It seemed a no-brainer and all went well until we came to claim the money.

Part of the attraction was that it was simple to claim, there was no extra form to be filled in. All we had to do was fill in our annual 'IACS form'[12] with the acreage grown, which is something we have to do anyway, and the money would be paid. The problem was that, although we were only being paid a fixed amount to grow the green manure in parts of fields each year, the first year I decided that I would plant the remaining parts of the fields at my expense so that all the land received the benefit of it. Richard, our land agent who came for an annual form filling visit, and I dutifully filled in the box to show what was growing in the field, as we were bound to do, and that was when the problem started.

The system accused me of trying to over-claim. I had one of those circular arguments with the young civil servant in charge:

'But you have over-claimed.'

'But I only truthfully stated the acreage we were growing and I can't over-claim because the contract says quite clearly that I will only be paid X.'

'Yes but you have over-claimed.'

Whatever. We will appeal and if that fails Richard assures me that his firm's professional indemnity insurance will cover the loss.[13]

Whatever the rights and wrongs of the EU's mad subsidy system, in the meantime, I am £20,000 out of pocket. We are bumping up

---

12. Actually now known as a Single Application Form (SAF) but still referred to by its previous name 'ayacks'.

13. Months later, following an appeal adjudicated by a civil servant in the same office, we did receive most of the money from the Department and Richard's firm decently made up the rest. It would have been scant comfort if we had gone under in the meantime.

against our overdraft limit. I had been relying on this money to pay bills and service the mortgage.

Whether I like it or not, change is being forced on us now. With trepidation I ring the bank managers. The first call is to the one who deals with our current account to request a meeting and plead with her to extend the overdraft 'while we carry out a major review of our farming operation. We are going to look at all the options. Nothing is off the table.' Her answer is reassuring. 'I'm sorry, there are a lot of others in the same boat. You'll have to join the back of the queue as far as a meeting goes. But don't worry, we'll extend the overdraft to give you space to work things out.'

The other call is to the bank manager who deals with our capital borrowings – a sizeable seven-figure mortgage that we took out when we bought most of the estate back in the late nineties and extra borrowings for all the improvements I have made. He is also encouraging when I stutter a request to convert our loans to interest-only while I try and work things out.

'I'll come and see you. Don't worry, it's not the bank's policy to foreclose unless we absolutely have to and we will do all we can to help.'

I start to feel a bit better about life. We don't have a plan but at least the immediate cashflow worries can be put to one side while I do some serious thinking.

CHAPTER 2

Spring Fever

It's nearing peak bird nesting time and John, our holiday cottage gardener/handyman-cum-volunteer gamekeeper reports that the Larsen traps are out. The traps are cages in which call birds – carrion crows or magpies – are placed to attract others who fall through the trap doors on top and can then be killed. It sounds cruel but nature is red in tooth and claw. Without us the crows and magpies would have no natural predators and would wreak havoc on the songbirds – which is why we do it; it does not serve any agricultural purpose for us to expend time and energy in this way, and although sometimes we do put down some pheasants, the corvids don't pose them any threat as the poults are too big for them to take when they arrive. We do it under a general licence to kill certain predator species.

Meanwhile there is uproar across the border in England as the licence has been withdrawn and the *Daily Telegraph* Comment Desk asks me to write about it:[14]

Classic FM listeners regularly vote for Vaughan Williams's masterpiece 'The Lark Ascending' as their favourite piece of music. It touches us more than any other piece of music perhaps because it mimics nature so closely and the liquid exuberance of the lark's singing, as he hovers above his nest, is for many of us the iconic sound of an English spring.

---

14. *Daily Telegraph,* 26 April 2019. This is the unedited version, the newspaper edited it slightly.

Perhaps that is why so many people – particularly rural people, as the lark is no metropolitan – are so bloody livid this morning, because the lark, already increasingly rare, has edged one step closer to extinction this week. Why? Because a highly paid BBC presenter has misused the power he has been given by our state broadcaster, paid for by millions of licence payers, to bully the government agency responsible for saying which birds can be controlled (and with the power to change the law without recourse to Parliament. Really?) And in the very week that farmers, those of us who genuinely care about preserving birds like the lark, are setting Larsen traps to control their number one enemy, the carrion crow, they are told that it is no longer legal. Just as the lark population is at its most vulnerable the muppets have declared open season on it for every crow and magpie in the country.

Natural England has bowed to a legal challenge from environmentalists led by Chris Packham, who call themselves Wild Justice, surely an oxymoron if ever there was one. Previously, the government issued a list of birds it was permissible to kill under general licence if they were causing damage. Under the old laws, they did not have to ask permission to kill the animals or record their deaths or the reason for shooting them. Now, Natural England has withdrawn all general licences while they 'work at pace to put in place over the next few weeks alternative measures to allow lawful control of these bird species to continue where necessary'. It draws attention to the baleful influence of the EU in our legal framework. The whole system of licensing is based in continental Roman law, which is antithetical to the presumption in English law that you can do something until it is specifically banned. We really are in *Alice Through the Looking Glass* territory.

Vaughan Williams's composition is based on George Meredith's poem of the same name, and Meredith could have been writing about Packham when he wrote:

> Unthinking save that he may give
> His voice the outlet, there to live
> Renew'd in endless notes of glee,
> So thirsty of his voice is he

He just can't resist it can he? Many will feel that Chris Packham has gone too far this time. Following on from his disgraceful accusation against the shooting fraternity in 2017 that the decline in the lapwing population was due to shooting (for which he later apologised) he has now caused untold damage to our songbird population by his clever-clogs court case, which focuses on a technicality in the way licences are issued and will not, in any case, lead to a permanent ban. There is a petition for the BBC to sack Chris Packham; if he survives this time stand by for a revolt as country folk refuse to pay their licence fees. This will be the final straw for many who were already feeling alienated by the il-liberal media's broadcasting of the Rousseauist agenda of Packham, Monbiot et al, which is based on dogma rather than science.

We need an honest debate, one that focuses on empirical evidence not emotion, one where we are allowed to hear the voices of those who care about the countryside because they care for the countryside 24/7. It is simply wrong to say that a laissez-faire attitude to our wildlife will provide a balance in nature – there are too many cats, grey squirrels and other predators let loose by man. The crow, which is omnivorous and can survive on carrion, is out of control, its numbers boosted by the food provided by roadkill and rubbish tips. There are very few apex predators left in the British countryside. Man needs to perform that role. It is time that the BBC got itself a new wildlife presenter and allowed a different message on conservation, before it is too late for the lark.

* * * *

Every year country people feel their way of life to be more under threat from the 'Neo-Roos'. Neo-Roo is an abbreviation for Neo-Rousseauist – I am minting a new expression here. They form a loose leftist alliance that in *Red Rag to a Bull* I called the Axis of Spite. They are sometimes called 'watermelons', a term popularised by my friend James Delingpole – green on the outside but red to the core. And in Scotland the term GRINOs – green in name only – is gaining currency. It includes politicians, civil servants, quangocrats, academics, scientists, animal rights activists, charities directors, journalists and BBC presenters and now, most visibly, Extinction Rebellion. Blair's victory in 1997 triggered the first assault – on fox hunting. At first it felt like traditional class war, a reflexive retaliation from Labour's post-industrial heartlands for the turmoil of the Thatcher years. But it soon became clear that the protagonists were middle class and metropolitan, and very well organised. Reasoned arguments by pro-hunting supporters on the Right like Roger Scruton and Charles Moore were outflanked by cleverly crafted tropes.

The overwhelming evidence that foxes – flight animals that have evolved through being chased in the wild – are either killed instantly by hounds or live to fight another day, the only form of fox control that avoids wounding and closely mimics natural selection, was simply buried by the Neo-Roos falsely persuading people that foxes are ripped apart by hounds while they are still alive – and that the *raison d'être* of hunting is to watch that. Hunting's genuine classlessness was cleverly redefined using identity politics. Its majority support in the countryside and the relatively small minority of anti-hunting activists were skilfully downplayed by a BBC determined to deploy equivalence with devastating effect. Even *The Archers* storylines were bent to the cause. As countryside organisations and campaigners have battled over the last two decades to defend the rural way of life armed with rational Enlightenment arguments, they have been outmanoeuvred by a post-modernist enemy that simply changes the truth if the truth doesn't fit their agenda.

The hunting ban was the first big concession made by Westminster to the Neo-Roos. But it didn't stop there. It became clear that this was something much bigger than the animal rights movement. It began to feel as if the countryside was now where the battle between Right and Left was to be fought in a way that it had not been at all in the twentieth century, and perhaps not since the 1840s. Slowly the penny dropped that the end of the Cold War had not seen communism vanquished, it had merely morphed back into an earlier form: Rousseauism. The ideology of the Swiss-French eighteenth-century philosopher provided the doctrinal underpinnings for the Jacobins during the French Revolution:

The first man who, having fenced in a piece of land, said 'This is mine', and found people naïve enough to believe him, that man was the true founder of civil society. From how many crimes, wars, and murders, from how many horrors and misfortunes might not any one have saved mankind, by pulling up the stakes, or filling up the ditch, and crying to his fellows: Beware of listening to this impostor; you are undone if you once forget that the fruits of the earth belong to us all, and the earth itself to nobody.

Rousseau, 1754

Rousseau's thinking is behind the Left's atavistic call to take the land back and its Counter-Enlightenment rejection of man's management of the environment in favour of a return to its primordial state. While the obvious failures of communism behind the Iron Curtain made it difficult to argue for seizing the commanding heights of the economy, environmentalism was an easy sell. Neo-Rousseauism with its heady mixture of climate change activism and anti-capitalist rebellion, and its romantic call to 're-wild' the land is what now captures the imagination of the disaffected. And its *casus belli* started to have far greater traction as concerns about climate change rose swiftly up the political agenda, pushed hard by Neo-Roos in the media.

Old-style Marxists found an easier way into politics through the Green Parties than via the Communist Party of Great Britain. In Scotland they have even reached power with the Scottish Greens (main policy: land reform) propping up the SNP-led coalition government. The proliferation of quangos set up by government to manage the countryside, partly to keep it compliant with ubiquitous EU environmental directives, played nicely into the hands of the Neo-Roos. The big charities were swiftly infiltrated and then leveraged their huge wealth from donations and bequests to weaponise issues that suited the cause. The Marxist idea of political correctness helped to close down opposition, sceptics were branded 'climate deniers'.

As the drumbeat grew louder it was easier to delineate the aims of Neo-Rousseauism. Reading *Communist Rules for Revolution*, a pamphlet circulating in Germany at the end of the First World War, it is striking how similar the Neo-Roo playbook is to classical Marxist doctrine:

1. *Corrupt the young*
2. *Get control of all means of publicity*
3. *Divide the people into hostile groups*
4. *Destroy the people's faith in their leaders*
5. *Always preach true democracy but seize power as fast and ruthlessly as possible*
6. *By encouraging government extravagance destroy its credit*
7. *Encourage civil disorder*
8. *By specious argument cause the breakdown of old moral virtues.*

Does this sound familiar? A quick leaf through school syllabi, particularly English, History and Geography, confirms that the young are being fed a Neo-Roo narrative and the Neo-Roos have found fresh converts in the teaching profession and thereby the Greta Thunberg generation. They have not seized 'all means of publicity', but they have certainly achieved suzerainty in the BBC. Stars like Chris Packham were promoted and others like Robin Page and David Bellamy were purged. And there has been a conspicuous attempt to 'divide the people into hostile groups' with environmentalists pitted against almost anyone who attempts to make a living in the countryside and 'othering' of farmers and particularly gamekeepers through naked identity politics. There is also a relentless attempt to 'destroy the people's faith in their leaders' by criticising the government for failing to tackle climate change. This is usually couched in preachy democratic argument but the 'ruthlessness' is exposed when the mask slips and leading Neo-Roos are seen orchestrating Extinction Rebellion protests and 'encouraging civil disorder'. And their leaders are now openly talking about going outside the democratic process and using 'non-electoral' means to advance the cause, Marxism again. Attempts to force the government to move to 'Net Zero' so fast that the economy collapses, and with it the whole capitalist system, can be seen as 'encouraging government extravagance to destroy its credit'. And the wokeism implicit in any Neo-Roo argument carefully 'breaks down old moral virtues by specious argument'.

Our relationship with the animal world is a key fault line between Neo-Roos and non-believers, and clearly a useful one for 'dividing the people into hostile groups'. The vegan movement is its most egregious example, animal rights is another. The RSPCA displayed naked Neo-Rousseauism by fighting against the badger cull, despite overwhelming evidence that TB is a cruel, debilitating disease, and that restoring a healthy badger population was supported by most rural vets. But the watermelon prize for charitable Neo-Rousseauism (an oxymoron) goes to the RSPB.

* * *

Driving back from Dumfries I see a roadblock ahead and cars being turned around by a young policewoman. I wind down the window and ask what the problem is.

'There's a tree down just this side of Shambellie.'

There is a backroad that would take me round avoiding the fallen tree but it is half a mile beyond her roadblock. The alternative is a seven-mile detour round by Beeswing. She doesn't sound local and maybe doesn't know this.

'You know if you move your roadblock just up the road you could direct the traffic round by Kirkconnel?'

She leans into the car with a look of disdain on her face.

'Are you telling me how to do my job?'

I bite my lip and reverse to turn around. The loss of Dumfries and Galloway Constabulary and its replacement by the centralised Police Scotland or, as it has written down the side of its vehicles, Poileas Alba in Gaelic, in case any time travellers from the twelfth century need translation, has not been accompanied by an ethos of serving the public.

* * *

A grey squirrel crossed the drive in front of me this morning. It is deeply depressing as we have just been told that the squirrel pox is now three miles away from us. I thought we were clear of them. The traps are out, and at least being in Scotland we are not affected by the insanity of the general licensing fiasco. But the worry is that our red squirrels will eventually catch it anyway from red-to-red contact. I think I know how General Percival felt when the Japanese had him surrounded in Singapore.

* * *

'Well done, Dad. That's six morons, four prats, three w***ers and one who wants to dip you in honey and feed you to *Charlie's Angels* – no only joking, he wants to shoot you.' It is lucky that my children are old enough to think it is funny, rather than very scary and upsetting when their father is engulfed in a Twitter storm. My computer has developed a rash from all the poison in the online abuse from Chris Packham's supporters after the *Daily Telegraph* article. If you can judge a man by his friends then I was surely right to call for the BBC to get itself another wildlife presenter. In my opinion he has allowed himself to compromise the BBC's covenant of impartiality. Worse than that, the tone of the abuse confirms my suspicions that he has stoked the fires of prejudice against farmers, and I don't see why the licence fee payers should continue to subsidise that. The Beeb's response seems to be that Packham is a freelance contractor so nothing to do with them, even though he benefits from broadcasting his opinions with the gold standard endorsement of the BBC's reputation for reporting the gospel truth. How much longer are licence fee payers going to put up with this?

The backlash from my article in the Torygraph exceeded all expectations. As the late, great Corporal Jones used to say, 'They don't like it up 'em.' I even received the ultimate badge of honour – being accused of being 'an Etonian landowner', as if that disqualifies me from public debate, by the High Priest of Self-Righteousness himself, George Monbiot in the *Guardian*. I later write to their letters page asking what Stowe- and Oxford-educated Monbiot has against Etonians. They have the good grace to publish it. He is not, as far as I know, a landowner; nor does he even live in the countryside, but that doesn't stop him telling those who do manage the land that they are doing it all wrong, repetitively.

In fairness to Chris Packham, he probably also has to put up with a fair amount of abuse, with which I sympathise. There is a news story showing an aggrieved Packham at the gate of his Hampshire mansion, on which some lunatic has hung a brace of dead crows.

But did they? In the murky world of environmental activism fake news is a recognised tactic. The story was suspiciously quick to break in the *Daily Mirror* and the BBC.

* * *

Birdwatching doesn't get any better than this. I am sitting watching the sun rise, listening to the dawn chorus in a hide in a secret location in the Pennines. I am watching a blackcock lek. Blackgame are very rare in most parts of the country, though they were present on heathland in every county in England once. I have never seen a lek except on a screen before and I am completely blown away by the raw excitement of the spectacle and the magical, primeval sounds – tschooks and a priapic, bubbling sound that seems to come from throat vibrations and becomes almost deafening as the male black grouse dance about, their wattles a livid red and their lyre-shaped white tail feathers going through elaborate turkey fan displays. Every now and again a greyhen flies in and all hell breaks loose as the males up their displays and fights break out before they mate. There must be well over twenty blackgame at the height of the party. It is nature's most exuberant orgy and the human voyeurs sit mesmerised in our tents. Then we walk across the moor, seeing a lapwing brood, curlew and golden plover nests and a healthy population of cock (red) grouse noisily patrolling their territories. I am bowled over by the care and expense that is being lavished on this fragile ecosystem, mostly through predator control, to maintain all this biodiversity. It is a tragedy that *Springwatch* presenters aren't there to beam this good news into the nation's living rooms.

It is always a privilege to be asked grouse shooting. But I am now realising that being asked to go 'lekking' is an even greater honour and for this moor owner, who wishes to remain nameless for obvious reasons, and probably for most others, the main motivation in owning a moor, which can be a financially ruinous business, is to conserve its wildlife. The shooting of a harvestable surplus of grouse, which may only happen three years out of five, is a bonus.

The Neo-Roo narrative focuses solely on the eight, admittedly very fortunate and mostly well-off people shooting the grouse each day, on perhaps only half a dozen days in early autumn, while ignoring all the benefits for an extraordinary array of wildlife from the keepering on the other 350 days of the year. It also ignores the legions of keepers and beaters earning around £50 to £100 for a day on the moors. A driven day on a big moor can provide work for over fifty people directly. On top of that there is extra employment in the hotels and shops. The familiar statistics wearily trotted out on these occasions, of the £2 billion contributed to the rural economy by shooting and the 76,000 full time equivalent jobs it supports, ignore the 'progressive' nature of a sport that transfers money from some of the wealthiest in society to some of the poorest in remote areas where there is little other employment.

Grouse keepers are probably the most traduced people in society. They can sit up all night protecting a hen harrier nest from foxes then be accused of killing the chicks when they are predated by a protected (but all too common) buzzard. The keepers themselves are trenchant about the criticism and were I to relate their comments in full you would probably have your vocabulary much enlarged. It always amuses me when I catch a rare glimpse of these unique moorland societies to see the great and the good treat these men with the deferential respect that they once accorded their colour sergeants at Sandhurst. Hard men, often with ebulliently disrespectful exteriors formed from a life on the fells, wary of outsiders but with a deep knowledge to pass on to those prepared to ask and listen. Their understanding of the natural world far surpasses that of the dogmatic young graduates in the Green quangocracy who nevertheless now control increasing areas of their work.

***

A book comes in for review.[15] I had not heard of Mary Colwell before but she is an excellent nature writer with a beautifully written voyage of discovery, following the death of her mother, as she sets out on 'a pilgrimage, an inner and outer journey' to walk across Ireland and then Britain in search of curlews and 'the

---

15. *Curlew Moon* by Mary Colwell (William Collins).

haunting sound of hiraeth, of love, loss, joy and sorrow that is the call of the curlew'.

Living on the Solway Firth in prime curlew wintering grounds, I share Colwell's love of this strange, primeval bird with its extraordinary vocal expression, and had been harbouring a gnawing anxiety for its well-being.

Winters on the Solway without the plangent cry of the curlew echoing across the mudflats do not bear thinking about. The book is hard-hitting in its call to action to save *Numenius arquata*, the Eurasian curlew, which is now officially our avian species most vulnerable to extinction.

Threaded through the lyrically evocative description of her walk across the wild, wet places of these islands to seek 'the bubbling song of courtship' are some uncompromising truths about 'the new moon bird's' decline, by 60 to 80 per cent since the 1980s. As a farmer, some of it makes uncomfortable reading. There is no doubting her analysis that the single biggest factor, below the heather line, is cutting silage too frequently during the summer months to allow curlews to breed successfully.

She visits wildlife groups to witness netting, ringing, fitting of satellite trackers and nest protection. She stops short of saying what many have argued: that you can have lots of badgers or lots of ground-nesting birds but not both. But she doesn't shy away from describing one project where cameras repeatedly filmed badgers destroying curlew nests. And she links the relative success of one Irish project to the enlistment of the local gun clubs to kill crows. Those who campaigned for the suspension of the general licence to kill corvids this spring should hang their heads in shame after reading this book. The curlews will have had a bad breeding season this year as a consequence.

Her honesty is refreshing. So many wildlife writers and television presenters are compromised by the Neo-Rousseauist agenda. It would have been easy and predictable to indulge in identity politics and write a book that vilifies landowners and, in particular, shooting interests. The curlew was, after all, a game bird until we stopped shooting them in the 1950s, although the Irish only stopped shooting them in 2012, and controversially it remains legitimate quarry in countries like France and Russia. But Colwell, despite describing herself as 'left-wing, vegetarian with vegan tendencies' is refreshingly objective in her analysis. After a visit to the Bolton Estate in Wensleydale she writes, 'The uncomfortable fact remains that I saw more curlews on grouse moors than anywhere else in the country' and, 'There is a general perception that protecting wildlife is soft and nurturing, but the reality on the ground is often raw and bloody. Some animals may have to die so that others can live.' This is a remarkable epiphany.

June 2019

Some good news: we seem to be winning the bitter battle to save the red squirrel here in the parish of Kirkbean – and I am crossing all my fingers and toes. All seemed lost when they confirmed cases of squirrel pox five miles away and at the same time we were seeing greys in our woods. I assumed that these were the reconnaissance troops of a tree rat army and that we would be powerless in the face of the grey blitzkrieg not far behind. We had seen it with Dutch elm disease, when our remote peninsula was one of the last places in the UK to succumb to it, and we watched impotently as our beautiful trees died one by one. Now it seemed that our cherished red squirrels would also soon be just a memory to tell our grandchildren.

But I had underestimated the Saving Scotland's Red Squirrels team at the Scottish Wildlife Trust. Within a couple of days of my SOS Stephanie and Steve pulled up in the yard. Stephanie is a charming Australian ecologist with a practical, down-to-earth approach to conserving wildlife. Steve had the menacing air of an assassin. Their vehicle was stuffed with traps and bags of special squirrel mixture. Stephanie reassured me that there had been squirrel pox breakdowns in Dumfriesshire before and they had been contained and healthy populations of reds restored.

With the sea on three sides and heather inland there are only so many ways they can reach us. So I mapped out a defensive plan with the two of them. They would augment our trapping efforts and clear our woods of greys then push out across neighbouring farms and hold the line several miles away from us either side along the coast so that our red squirrel colony could hold out until the war is won nationally. We toured the parish looking for likely habitat and all the neighbours were keen to help. One month later and there have been nine greys killed in our woods and plenty of healthy reds trapped and released. The bridgehead is now being pushed out.

By chance I played golf at Woburn in Bedfordshire this week. The family home of the Duke of Bedford is the epicentre of the grey's evil empire as the 11th Duke was the muppet who introduced the American invader in the first place in 1890. The golf was terrific but whereas playing at one of our local courses in Galloway is an ornithological treat at this time of year, in thirty-six holes I only saw three songbirds. Every fairway had crows pecking around in the middle of it and there were grey squirrels everywhere one looked. Above, red kites soared, ready to pounce on any fledgling unwise enough to hop out into the open. Several times I mistook plundered eggs for my ball and we saw hen pheasants and ducks that should have been sitting on eggs or with broods but had clearly lost the chance to breed this spring. It was heartbreaking to see and for our team of Scottish farmers it was an apocalyptic vision of what happens when nature is allowed to get out of balance.

As it happened, our team captain was one of the trustees of the Red Squirrel Survival Trust and he was able to report that there is a national campaign, initiated by the Prince of Wales, to rid the country of grey squirrels. This will be done by clearing whole areas of greys by feeding contraceptives during the breeding season for several years as well as trapping, then re-introducing the reds. There is also a programme to develop squirrel pox vaccine. Landowners in Cornwall, led by HRH the Duke, have started to roll up the grey population from the west and re-establish reds as they go.

It is going to be an uphill struggle. There are moves by Neo-Roos to have the grey reclassified as a native species. And urban households, for whom grey squirrels on their bird tables are the only wildlife they see, will not take kindly to seeing them eradicated. But we must prevail.

CHAPTER 3

Back from the Brink

Bank managers are often characterised as people who lend you umbrellas while the sun is shining then ask for them back when it starts raining, and we have had some shockers during my farming career. But John, when he comes to lunch, is one of the best. Agricultural bank mangers undergo a rigorous training and then spend their working lives looking at different farms and, as importantly, their balance sheets. They generally know what works and what doesn't and they have seen it all before. I have learnt to value their advice as much, if not more than agricultural consultants, particularly as it is free.

It is a real help to run through the various options with him. I have discounted completely the idea of letting the land out. The Scottish government could never be trusted not to legislate in favour of tenants so that we might never get the land back again, even if the tenants turn out to be no good. Worse, we might even be compelled to sell them the farm. I have had enough bad harvests to know that arable farming in the west would pose as much of an existential threat to us as beef farming has. And doubling down on beef is likewise out of the question, as is covering the land in thousands of sheep.

We could do annual grazing lets, but if our fears for the beef industry materialise then we might struggle to find graziers to pay enough rent. There is demand for 'cutting ground': producing silage for other farmers either to feed cattle or to put into 'concrete cows' – the anaerobic digesters springing up around the county.

But this would destroy much of our wildlife and rob us of fertility. We can't put in an anaerobic digester ourselves as we haven't a good enough grid connection.

'It keeps coming back to dairying. I know that's what our land should be in really. It's what it was in my grandfather's time. It was only the artificialities of being in the EU that led us into other things. But would the bank lend us the money to convert?'

John is wise and measured in his response. The bank's view is that dairying is possibly less vulnerable to Brexit than other sectors, partly because most dairy products are perishable commodities. He points out that we are slap bang in the middle of one of the biggest milk fields in the world, stretching from Glasgow down to Devon and from Stranraer across to the Pennines, so we are always going to be close to a processor and we have the soil and climate to grow lots of grass. He advises me to look into it seriously and find a partner with a proven track record.

'It's astonishing the difference between dairy farms. I can have one client losing money hand over fist on one farm and another one next door who is doing it differently and making a lot of money. It's all down to the type of operation they run.'

It is a worry knowing that in order to dig ourselves out of debt we have to take on a lot more debt but the more I think about it, the more I know that is what we must do. From the scraps of training I received at business school I remember that it is not the absolute level of debt that matters but the gearing – the ratio of debt to assets and the margin by which the borrowing costs are covered. If we can turn the estate back into a profitable dairying business that is worth more and cover our borrowings more comfortably our gearing would actually be reduced. I am mindful of two similar situations where farmers have suddenly been exposed to free trade and no subsidies after a period of protectionism. First the period after the First World War when A. G. Street wrote his classic book *Farmer's Glory* about how he survived by outdoor dairying and setting up a milk round. And secondly when New Zealand removed their farm subsidies in 1984 and their dairy farmers became very efficient and thrived. I have to think like a Kiwi.

* * *

Another book arrives for review.[16]

This is classic Neo-Roo – a potted history of the Cold War: 'Ronald Reagan reversed years of détente with the Soviet Union and began calling it the Evil Empire ... with the sudden end of the Cold War, the cruise missiles left Greenham ... the perimeter fence was torn down and common rights restored.' This comical interpretation of events aside, it is evident that whereas for millions freed from the tyranny of socialism behind the Iron Curtain, and indeed for the dog walkers of West Berkshire, this was palpably 'a good thing', for Guy Shrubsole it obviously came as something of a disappointment. Brought up to demonstrate alongside the Greenham women by his CND-supporting mother, suddenly he finds himself without a cause.

And there you have it in a nutshell. It is a commonplace to explain the violent upsurge in environmentalism by 'the failure of capitalism' in 2008. But, as this book demonstrates, its roots lie much deeper.

He begins the book with a quote from Norman MacCaig:

Who possesses this landscape?
The man who bought it or I who am possessed by it?

My answer to that question would be both of us. I see the land as having many stakeholders: the little platoon of family and estate workers who lovingly care for it, the tourists and ramblers who come and enjoy it (and, unfortunately, the small army of civil servants who inspect it and regulate it with ever-increasing frequency). Shrubsole illustrates that if the sixty million acres of the UK were shared equally amongst the population we would each have an allotment of one acre. Beyond this superficially attractive proposition it is a dystopian vision, if you think about it, as the landscape might come to resemble, well, the Greenham Women's Peace Camp and, shorn of the economies of scale, our agriculture would soon revert to subsistence levels.

---

16. *Who Owns England? How We Lost our Green and Pleasant Land & How to Take it Back* by Guy Shrubsole.

The book is useful, not in the way intended as a desirable blueprint for land reform, but as a chilling insight into the mindset of a Neo-Roo.

* * *

My depression about the state of beef farming is not helped by watching the ten o'clock news. BBC Science Editor David Shukman shows the graphic of a football pitch-sized area of primary jungle in the Amazon basin being chopped down to make way for beef cattle. Apparently one goes every minute and, to emphasise the point, a day's worth of football pitches zig-zags across the screen. Then the camera switches to a field of lop-eared cattle grazing on the edge of the jungle. I bet it will knock several pence off the British beef price even though there is no comparison, in environmental terms, between grass-reared British beef and Brazilian beef that has released tons of carbon into the atmosphere before it has even been turned out to graze amidst the smouldering fires of ex-jungle.

It is already gloom and doom because farm-gate prices are at an all-time low. It probably has more to do with pre-Brexit fears leading to stockpiling in the supply chain than the growth of veganism. But sharing my concerns with the neighbours it is clear that in the pit of all our stomachs is the fear that this is the shape of things to come because consumers, and therefore politicians, will turn their backs on the meat industry, driven by fears of climate change. Here in south-west Scotland the agricultural economy is based on cattle and sheep because the soil and the climate dictate that grass is the best crop to grow and often the only possible one. Shukman (whose family are Dumfriesshire landowners) must know that. Without the ability to grow meat and milk the only solution is to plant the land up with trees and go and do something else for thirty years while they grow. Rural depopulation is already happening and it is being likened to the Highland Clearances.

'Ah,' I hear you say, 'but what about methane?' Well, yes, methane is a greenhouse gas, one created by ruminants, amongst other sources (rice paddies), then recycled back into the soil by breaking down into carbon dioxide. To put it crudely, every time a cow burps, the burp[17] from a cow years ago disappears simultaneously by breaking down into carbon dioxide and being reabsorbed by plants. It is part of a natural cycle that has been running since God was a boy.[18]

Nor do the cattle create the methane. It comes from the vegetation decomposing in their rumens. That vegetation would decompose anyway if left to rot naturally or if fed to vegans, since that seems to be the preferred option.

Removing the source of methane might seem like a clever short-term fix, but would end in catastrophe as it would destroy the soil's potential for sequestering carbon. Better by far to make a virtue of necessity. The dairy industry is starting to harvest biogas from slurry, which is already being burnt to create electricity on British farms. Arla, the UK's biggest dairy processor, estimates one cow can produce enough to power three households annually, so 16 per cent of British homes could be slurry powered. I have a friend with a (legal, medicinal, promise) cannabis farm. He plans to burn methane from dairy farms that will heat his greenhouses and create carbon dioxide to pump in to make the plants grow faster. And if you are still terrified by methane, there will soon be anti-flatulence feed additives for cattle made from seaweed that will reduce emissions by up to 82 per cent.

Then there is the muck that would need to be replaced by artificial fertilisers, which are one of the biggest causes of greenhouse gases through the Haber-Bosch process, the leather substitutes, glue, even jelly babies. The faux environmentalists never seem to mention the other side of the equation.

If we rid the British countryside of the humble British cowpat we would remove the elemental building block of much of our indigenous insect life, and consequently birdlife. Desperate to uncover the truth about the good that cows do, I consult one of the

---

17. For complex biological reasons, cow burps comprise mostly methane whilst farts are mainly carbon dioxide.

18. Fortunately a global team of Intergovernmental Panel on Climate Change researchers based at Oxford, led by Professor Myles Allen, is making some headway in re-evaluating the effects of methane. But it will be uphill work persuading the Neo-Roos.

most eminent ornithologists in the country, Professor Ian Newton, a Fellow of the Royal Society and a recipient of the RSPB's coveted gold medal. He confirms that the reason for the steep decline in insect numbers is partly the lack of cows in parts of the British countryside as mixed farms have become all-arable, and partly the use of ivermectin wormers that kill all the bugs in cowpats. He kindly sends me an article[19] he has written:

> In a pioneering study, Lawrence (1954)[20] found that, on average, each cowpat produced about 1,000 developing insects. Each animal deposited 7–10 pats per day, but some were destroyed by trampling or in other ways, so he assumed six suitable pats per day. This was equivalent to 6000 insects per day, or nearly 2.2 million insects per year (mostly flies) for each beast kept outside year-round (these estimates are not, of course, applicable to dung stored as muck-heaps or slurry). Accepting seasonal and other variations, Lawrence went on to estimate the total annual production of insect biomass from the dung of each cow or bullock kept on pasture. He concluded that 'a cow leaves in its faeces enough food material in a year to support an insect population, mostly dipterous larvae, equal to at least one-fifth of its own weight.' Not all insects that used the dung could be included in his calculation, so for this and other reasons, his estimate should be regarded as minimal. It also excludes worms of various kinds, which are also eaten by birds. But as a rough guide, we could say that, in five years, each cow or bullock kept outside on pasture can produce its own weight in dung insects.

Why do we not hear more about this from the BBC? It seems madness to reject a juicy slab of steak grown in the green grass fields of Galloway in favour of a measly soya substitute grown on what was once beautiful rainforest in South America, but there is a steady drip of advice from our state broadcaster saying that we should eat less meat. It forms a depressing backdrop to

---

19. Ian Newton, *British Birds* 111, November 2018, 636–638.

20. Lawrence, B. R., 1954. The larval inhabitants of cowpats. J. Anim. Ecol. 23: 234–260.

my deliberations. Meat is an important by-product of the dairy industry; the only saving grace is that vegetarianism actually helps dairy farmers, as vegetarians eat a lot more cheese.

August 2019

At last, there is a chink of light at the end of the long, dark tunnel our farming journey is taking us on. I have been quietly sounding out dairy farmers and today Sheri and I are having lunch with some potential partners.

Richard Beattie and Brendan Muldowney have come to make us a proposal. They have built up a successful dairying business called *Farming Partners*, based on an idea that comes from New Zealand, where both of them have gained valuable experience, that of 'share milking'. The basic concept is that we would put in the land and the infrastructure for a 650-cow dairy and they would put in the cows and their expertise and we would then run a dairy together producing three million litres of milk each year, sharing the income and expenditure equally. Their model has been shown to work without the need for farm subsidies and they have a proven track record.

We have had a look around the estate and they think that their operation will work subject to a number of 'known unknowns' and probably a few unknown ones as well. Theirs is an all-grazing technique where the cows are run in mobs of about 300 which graze small paddocks for twelve hours at a time before walking along cow tracks, which we need to construct, to the parlour to be milked. The cows are small Jersey-Friesian crosses and they produce high butterfat milk for which there is a premium.

For me it ticks a lot of boxes. The alternative would be a 'zero-grazing' operation where the cows never go outside and the land is silaged four times each summer to provide the feed, or perhaps put into maize, which in this part of the world needs to be grown under plastic. Its exponents would say that is more efficient, as their large bony Holstein cows produce more milk, mostly 'white water', than Richard and Brendan's little Jersey crosses. But it isn't how I want to farm.

Importantly it gets the estate further down the path I have been trying to take it – that of regenerative agriculture, which focuses on building life in the soil. Strangely most of us know more about life on Mars than we know about the soil beneath our feet. A teaspoon of (healthy) soil contains more living organisms than there are humans on earth, and around 10,000 species. We are only just beginning to understand what goes on in the 'rhizosphere' – and the more we discover the more we realise that ploughing the land to grow 'plant-based foods' is catastrophically bad, the result of a wrong turning in agriculture around 5,500 years ago. And what is good and increasingly necessary is planning all our farming systems, particularly arable ones, around cattle grazing, defecating and trampling vegetation to revitalise our depleted topsoils so that they have the right mix of bacteria and fungi and can absorb carbon dioxide through plant growth and put carbon back under the sward, where it belongs; along with nitrogen craftily harvested from the atmosphere by legumes such as clover. That is the thesis of the regenerative agriculture movement led by Zimbabwean ecologist Alan Savory and North Dakota rancher Gabe Brown. If these two men are not household names in the UK it is only because the BBC has failed in its mission to inform and educate us – perhaps because it is promoting a different narrative – and you have to flick to Netflix to watch the brilliant documentary *Kiss the Ground*.

I also like their share-farming business model because of its implicit fairness and the way it incentivises their young managers, who are able to build their own equity in the business by owning a share of the cattle. The conventional landlord and tenant model is utterly broken, though you won't get a politician to admit that. Its problem is that someone always has to lose: the rent is either set too low, and the landlord loses, or too high and the tenant does. 'Share milking' is completely fair to both sides.

And finally, although the sums involved are scary, it is a relatively affordable option. They only need an outdoor parlour. There is no requirement for massive buildings or robot milkers. All of the cattle sheds and handling facilities I have built for the beef operation can be converted relatively easily. If I were to set up a dairy operation from a green field site including the herd, I would be looking at an investment of around £3 million, which would be well beyond my borrowing powers. This would get us back into dairying for much less than that.

We provisionally agree to work together subject to our being able to agree terms and surmount a huge number of hurdles. We don't know what it will cost or if we can borrow the money. We don't know if we will get planning permission. We don't know if we will get a milk contract. We don't know if we can find enough stone on the estate to build all the tracks we need. But we have a plan. It feels as if a huge burden has been lifted from my shoulders.

* * *

The curlew's searing cry across the mudflats has been a disturbing feature of life on the Solway Firth this summer. I say disturbing because, whilst I am delighted to hear them here when there is an 'r' in the month, curlews are supposed to be safely tucked up with their young in their upland breeding grounds in June and July. It is possible, putting an optimistic spin on it, that these are birds that are leaving their mates in charge and flying off to feed before returning inland. We know that they do that, sometimes travelling long distances for a seafood feast. But the numbers of them and the fact that we are seeing quite a few pairs leads to the depressing conclusion that these are birds that have tried and failed to raise a brood and have cut their losses and decided to head for the seaside earlier. The weeks of the ban on Larsen trapping on the English side of the border this spring will not have helped and predation by crows that might otherwise have been controlled may even have been a deciding factor.

On a more cheerful note, I know of one grouse moor restoration project where previously there had been no curlews and this spring they have had fourteen successful broods. Curlews live several decades and only need to raise a handful to replace themselves so this gives one hope that all is not lost for one of our most iconic and endangered waders. Sadly the location has to remain anonymous for obvious reasons. They don't want a dead raptor planted on them in a depressingly predictable retaliation for publicising a brilliant conservation story paid for by the anticipated revenue from grouse shooting.

September 2019

A glorious September morning of driven teal and snipe will stick in the mind, especially one drive where we had to look hard for the snipe through an aerial display of swallows and sand martins flexing their wing muscles before the long journey south. It will also stick in my mind for what I learnt that day. My friend the amateur keeper looked pale. At least I thought he looked pale; it's quite hard to tell under the blotched walrus hide. 'I had a visit from the RSPB,' he confided. 'Two of them, outside my release pen in broad daylight. They said they were there to do an inspection and then proceeded to search my pick-up.' The aforementioned vehicle is famed across the county for the depth of detritus covering every surface so I was not surprised when he finished on a satisfied note. 'They didn't find anything.' He told me this with the pained tone of a downtrodden citizen of a communist state where these sorts of infringements of civil liberties are commonplace.

I have never understood why so much attention is paid to the peregrinations of the RSPB. To anyone working on the land they are a discredited organisation, widely ridiculed for the way that they take huge sums of money from donations and grants and manage to deliver decreases in bird populations, when compared to other landowners' ground, on their own reserves like Geltsdale and Lake Vyrnwy. So omnipotent have they become, I was unsure myself whether the Revolutionary Sect for the Persecution of Barons had indeed been awarded some new powers. It gave me an uneasy feeling so when I got home I checked with the in-house legal team of a respected farming body to which I belong, who said that mine was not the first enquiry along these lines. But no, the two RSPB employees (if that is who they were. Maybe they were volunteers, or perhaps imposters, there is only their say-so to go on) did not have authority to search my friend's car and were almost certainly acting unlawfully, contrary to the European Convention on Human Rights. In fact, despite Scotland's wide-ranging permissions for access to the countryside they were almost certainly trespassing, as you may access land for leisure purposes but not to carry out a commercial activity for an employer, and an inspection on behalf of a charity would fall under that category.

How have we arrived at this situation? From the allegations being made we now appear to have an organisation established by Royal Charter – as distinct from being awarded statutory powers by parliament, which is an important distinction – taking it upon itself to act as if it were the police force. Their justification seems to be the same as for vigilantes everywhere: there aren't enough police to go around so something must be done. But of course they can never be a proxy for the police because crucially the police do not take sides in any debate and are scrupulously neutral. Policemen carry identification and need permission to search premises. And there is a robust complaints procedure to safeguard the citizen against over-zealous or corrupt policemen.

Lifting the stone I find several cases where judges have had to rule on the admissibility of evidence that has been gained by supporters of animal rights organisations in clear acts of trespass. Covert cameras have been placed on private land without permission. And there are allegations that one organisation, has been releasing golden eagles with trackers on in the hope of some hothead shooting them. Entrapment is itself of dubious legality but, if true, then this is a criminal activity. The golden eagle chicks will almost certainly have been taken from the nests of wild birds, something that can only be done under licence as there are strict protocols to be observed, which include leaving one chick and rearing the birds without them ever seeing a human.

We all know that there has been wildlife crime and that is wrong even if, as many of us believe, the law regarding the protection of certain species requires amendment. But the rule of law is much more important and the animal rights lobby is not above it. And our legislators should clearly define the powers of the RSPB and others and review their royal status if they have brought the law into disrepute by acting unlawfully. Nothing is more important than liberty and seventy-eight-year-olds should not live in fear of Neo-Rousseauist vigilantes.

CHAPTER 4

Hope

Whatever the weather, Mother Nature just gets on with the job. Heartbreaking sight of the year is a hen pheasant crossing the road on 7 September with seven-day-old chicks. She must have been sitting all through the recent wet weather to bring them into the world at a time when their chances of survival must be as statistically close to zero as possible. I like to think that she has already reared one or even two broods successfully but she has very likely experienced the heartbreak of losing them to the crows. A happier sight is a nest of swallows on the point of fledging. There is a small chance that they might make it to the Eastern Transvaal for Christmas given a fair wind, and they are definitely a third brood. It seems to have been a bumper year for frogs, toads and wasps, whatever that means. The latter have presented a particular challenge by a marked tendency to take up residence in our holiday cottage roofs. There is something consoling about the plagues in the Old Testament, which show that these freak natural population booms are nothing new.

Maybe undertakers remember particularly mild winters when an abnormal number of old people escaped the Grim Reaper. And I have known teachers shudder when recalling a particularly recalcitrant year group. But more than perhaps any other profession, except possibly stockbrokers, farmers carry memories of bad years with them for all time. And years are defined by harvests in arable farming households. In a wet August the farming press carries letters along the lines of, 'It's bad but not as bad as 19** when we got the combine bogged and couldn't dig it out until the following May.' And readers vie to outdo each other with heart-wringing tales of woe, contributing to the general anxiety. The harvest of 1985 nearly destroyed my father and he was still referring to it thirty years later. I still have nightmares about 2012 when the wheat went black and the grain shook out in the field so that, on the rare days when it wasn't raining, when you walked in the fields it sounded like the patter of a shower, the sound of money being lost. Contractors had a bad time of it with farmers pleading with them to come. It didn't help that the grain never hardened; it went from soft and 'too soon to cut' to sprouting in the ear in a matter of hours. Rubbing it between the palms of my hands left tell-tale green strands of wheat grass that cut me to the quick. We felt forsaken that year. The final nail in the coffin was the announcement of the harvest festival service. I had to point out to the minister that many of his parishioners had yet to start their harvest let alone finish it. But it went ahead anyhow, despite there being little to thank the Almighty for, another rip in the fabric joining that particular incumbent to his congregation.

So when the monsoon hit this August, and we could almost smell the Caribbean spices on the wind, there was a sinking feeling of here-we-go-again. It was 'catchy' all the way through to the day we cut the last field of spring wheat on 14 September. As with other wet years, too many in the last decade to count comfortably (which may support the theories of the climate change doomsters). I felt like a boxer hunched on the ropes, helpless to do anything about the onslaught as angry clouds poured blows on me. The advent of modern weather forecasting has only made things worse. At least our forebears could console themselves with misplaced optimism. They could *think* that it might be a fine day tomorrow, blissfully ignorant of the BBC's ten-day forecast on the computer showing rain every day backed up by lurid blue and green images heading across the Atlantic.

Harvest 2019 will remain embodied in grey hairs and wrinkles, and in ruts in fields that may take years to shift. But maybe someone is trying to tell me something. Having a dairy back on Arbigland would at least mean that we would no longer be at the mercy of bad harvests.

October 2019

Richard and Brendan have produced a budget and the bank are looking at it. Whether we can get the dairy off the ground or not, I have decided to press on and sell the beef herd, partly to keep the cashflow going and partly to cut down on our wintering costs. To spread the risk I am selling the cows with the calves at foot in tranches over the next year, so we will receive an average price rather than the going rate on any particular day. The price of breeding cattle is on the floor but still well above where it was when I established the herd in the wake of BSE and foot-and-mouth, so I can at least console myself that I have increased our capital, even if we have made losses recently. Every penny counts now as we plan the dairy development project.

One of the hardest things for me has been worrying about Davie and Graham. Their jobs hang in the balance, and along with them possibly their tied cottages. I have warned them that there is likely to be change but that I don't yet know what that is going to be. Their stress has been all too obvious as well. I feel guilty that I have let them down by pursuing a business model that is heading into a dead end.

I have dreaded telling them but it is a relief for us all when I let them know that I am selling the herd come what may and attempting to get back into dairying but whatever happens I will see them right. They know that while their wages have risen over 50 per cent over the years the price of beef has stood still and so it comes as no surprise.

* * *

The hunting gene is still present in us all and, judging by their television programmes, even animal rights activists are thrilled by the spectacle of one wild animal killing another. I have never been able to fathom the irony that no one ever objects to a naturalist filming a pack of wolves hunting a fox in the wild. Yet when a pack of hounds does the same thing with a small amount of human agency it attracts the opprobrium of the Axis of Spite. Falconry never seems to excite the same prejudice. During the campaign to save hunting from the class hatred of the Blair government I remember regretting that the gritty northern writer Barry Hines had not found his passion in the local hunt kennels rather than through the adoption of a kestrel. The similarities between falconry and fox hunting are obvious and ethically indistinguishable. Happily, the kinetic excitement of watching a falconer's hawk stoop on a partridge is something that is still a matter of individual conscience and has not been banned by the Neo-Roos.

Like the hero of the film *Kes*, our electrician Greg Hutchings was bitten by the falconry bug as a teenager:

> I was obsessed with my airgun and my bow and arrow and keen on birds of prey. I must have been twelve years old when I climbed a tree to see a sparrowhawk's nest with five chicks in it. Not long afterwards there was an *Open Space* TV programme about falconry and I was hooked. I read seven books about falconry on the trot, Barry Hines's *A Kestrel for a Knave* (the book on which *Kes* is based) among them. My dad was delighted as I hadn't been that keen on reading until then! I soon became an 'austringer', flying a buzzard off the fist and then, like many falconers, progressed to train sparrowhawks and to flying a goshawk when I was nineteen. Finally, in my late twenties, I moved onto flying 'long wings' such as peregrines.

Greg and I are sitting in our kitchen over a mug of coffee with our new friend, wildlife photographer Duncan Ireland. We are on a mission. Outside on his block, hooded and ready to go is Kenzie, Greg's fifth season peregrine named after one of Greg's heroes, the wildlife artist and Lincolnshire wildfowler Mackenzie Thorpe. Duncan and I are struck by Greg's enthusiasm as he talks about how he bred Kenzie from a wild female that had been handed into a wildlife reserve with a broken wing and fathered by a bird of Greg's, a retired twelve-year-old named Frodo. Kenzie was originally entered to duck but after a tussle with a big mallard drake several times his weight his confidence was shaken and he now flies against hen pheasants, grouse, partridges and teal.

Weight is taken as seriously in falconry as in boxing world title fights. Kenzie weighs in at 1 lb 4 oz. His flying weight varies according to the time of year. Birds have inbuilt barometers that tell them to lay down fat in anticipation of bad weather. Part of the skill is in ensuring that the bird is fed to perfection. Greg doesn't like to keep his birds too lean but if they are overfed they won't hunt.

Outside, the wind is shaking the first of the autumn leaves from the sycamores along the shore and we watch anxiously for it to drop. The maximum wind speed Greg will fly his bird in is 25 mph, as beyond that it will struggle to keep position. Today's expedition has been months in planning. For two years Greg has been putting down grey partridges here in the hope that one day he will witness Kenzie hurtle out of the sky and take one. In contrast to the thousands of redlegs released in Dumfries and Galloway, the native bird is all but extinct and Greg's conservation efforts have been a happy by-product of his falconry. Grouse are his quarry of choice and there is one heathery outcrop not far from here where a small stock thrives, largely thanks to Greg's keepering for falconry.

A couple of years ago he asked me if I would mind if he put down some grey partridges in some promising looking fields of ours. The answer was, 'Is the Pope a Catholic? But why?' He explained patiently, as if to a child, that partridges are the ideal size for his falcon and the French species or 'redlegs' are no use at all as they run[21] rather than clamp down to a point; they have to be the English variety. So he has worked tirelessly to try to re-introduce grey partridges on several farms in the district, all for the exciting prospect in his mind's eye of a covey exploding across the stubble and the lightning strike of his peregrine. There has been a lot of airtime devoted recently to our opponents' attempts to ban the release of 'non-native' partridges. But would they applaud Greg's valiant, and hitherto only partially rewarded, attempts to save the native partridge? Not if it involves killing I suspect.

The partridges that Greg puts down are bantam-reared. He buys them at five weeks and they are then grown on in Greg's garden on a diet of wild bird seed to accustom them to foraging in the wild. Then at eight to ten weeks they are gradually released from a pen on the edge of one of our stubbles where there is a thick beetle

---

21. Much like the French army, it is often opined.

bank of white grass, in such a way that a covey is then hefted onto part of the farm where there is the best possible habitat. Greg has been helped by his sons Oliver (twelve) and Lewis (ten), whose enthusiasm is only matched by Greg's.

Our old game books show that in the nineteenth century grey partridges were plentiful here but modern agriculture and an increase in predators had wiped them out by the 1970s. We have been allowing the hedges to thicken up and our seed rich 'green manure' crops have been ideal cover as well as providing plenty of natural food for birds. So far there have been some successes with partridges surviving through to the following year, although we have yet to see a wild brood. We have not been shooting them to allow their numbers to build up. Today we are going to see whether all Greg's hard work is going to be rewarded.

Greg talks us through the tactics. The plan centres on Hattie, his German short-haired pointer in whom Kenzie has developed confidence after a number of successful hunts together. Hattie has been known to hold a point for forty-five minutes while Greg has moved into position. Falconry pointers quarter more widely than those trained to the gun and from further away. Whereas shooting grouse over pointers requires the dog to get the guns to within forty yards of the flushing point, a falcon can strike from much further away. Suddenly I feel as I felt in the army when my narrow horizons as an infantryman were opened up to the possibilities of air support.

Though I had observed it happening often enough, it had never fully occurred to me that one of the main reasons why birds take off into wind is to blunt the attack of a raptor. What falconers try to do is 'head the point' by moving round beyond the quarry so that when it is flushed it will see them turn and fly downwind to assist the falcon's stoop with the wind behind.

The wind drops and we finish our coffee and head out, eager for action. Pulling up in a gateway by a likely stubble we have a chance to have a good look at Kenzie. I am struck, as always, by the neatness of the bird and how sweet it looks for a killer with its kind black eye, in contrast to the pathological yellow eyed stare of a goshawk. Nature has reached peak economy and perfection in the peregrine. The male is two thirds of the size of a female because evolution has worked out that it is better for them to hunt different species, though often they will work together with the tiercel lifting game for the female. We have wild peregrines here and the first time Greg came with Kenzie he was chased off by a wild bird.

Greg talks us through the kit.

'These are the jesses that we use to attach the bird. That's the origin of the term "big jessie" – an inexperienced falconer would need longer jesses.'

The hood and the jesses, with some clever modifications, are largely unchanged from when man first domesticated falcons. What has changed dramatically in the digital age is the ability to track the birds. The falconer used to rely on a bell attached to the bird to help him find it again if it flew off. Now a small antenna can be attached to the bird. The technology has been developed from military GPS technology by an American company, Marshall Electronics.

Greg attaches a small antenna to Kenzie's leg. Astonishingly it doesn't seem to affect his performance. What it does do, once Greg's phone picks up the signal, is send his exact location, which can save hours of anxious searching. Cleverer still, it has an altimeter and speedometer. Greg now knows how high his bird has flown and the speed of the stoop. His record is 161 mph from a height of 1,453 feet. Greg lifts Kenzie on his glove and we are all set.

As we enter the field a sparrowhawk flies low across the stubble as if it knows to make itself scarce. Hattie bounds expectantly, eyes on her master, waiting for the off. Greg sends her out, quartering across the stubble into wind until she spins suddenly onto point. Greg checks her with two blasts on his whistle and quickly removes the hood. Kenzie turns his head to stare intently in every direction.

'You see that? He is photographing his surroundings so that he can memorise them.'

Moments later the bird takes off and we watch in awe as he climbs and circles until he is a speck against the clouds, bringing the whole parish within his range. Duncan goes into paparazzi mode and snaps away frenetically. Watching through my binoculars I am fooled by the scale at a distance and think how like a swallow he looks as he gracefully soars and glides, cutting into wind to gain height above Greg. Other birds react to a killer on the loose, three fields away a cloud of pigeons lifts off a stubble and in the next one rooks and jackdaws wheel and swoop on the breeze. I worry that Kenzie may kill something out of view but his loyalty to Greg and Hattie draws him back so that he is soon 'waiting on' above us.

Duncan and I move off to flank a rough bit of cover and allow Greg to head Hattie's point when suddenly a hen pheasant is bumped and takes off not far from us and whirrs away into wind. We are watching it intently as it disappears over a knowe[22] in the next field when out of the corner of my eye I see a flash of slate-blue wings, Kenzie in full pursuit. He closes on the pheasant with frightening speed, lifts, stoops then ... nothing, the two disappear from view. Disappointment is short-lived as Kenzie quickly remounts to a position directly above Greg, who sends Hattie into flush and a covey of partridges breaks. The next few seconds are a blur of acrobatics as Kenzie stoops hard and fast, accelerating into the covey. He appears to miss one bird passing underneath him by inches then throws up and locks on to another. He courses the bird downwind then there is a puff of feathers as he seizes it in his talons. Success.

There is a moment in fox hunting, savoured by those of us old enough and lucky enough to remember, when hounds have killed their fox in the open and the pack gets its just rewards and breaks up the carcase – an entirely natural action, and one without cruelty as death has been instantaneous – and the field gathers around them, flushed with the thrill of the chase and feeling more alive and at one with nature than ever. So it is with falconry as we close in with Kenzie and his kill and watch as he eats the head and neck of the partridge until Greg gently replaces it with a pigeon breast and puts his share in his game bag, a ritual transaction that is millennia old. Duncan looks at his camera and confirms that he has captured it all. I love it when a plan comes together.

***

---

22. Local word for a hillock.

The sheep arrived today. Perhaps for the last time, as sheep won't form part of the new regime as the dairy cows will be out grazing until November and then again in February and the grass needs to recover in between. It is a bittersweet moment. The estate becomes a gentler place when our fields are 'where sheep may safely graze'. On the other hand Bach's cantata does not allow for the violent exercise they give the shepherd (me) when they escape or get caught up in wire, or selfishly die so that they require dragging to the knacker man's pick-up point.

The income from wintering 700 Romney hoggs (female lambs under a year old) helps to keep the bank manager happy. They play a vital role in managing my grassland organically by cleansing the pasture of the intestinal worms that prey on the cattle in the summer and they also control the poisonous ragwort so that I don't need to spray it. Their owner Marcus Maxwell is a former Sheep Farmer of the Year and an efficient, outward-looking farmer with interests in New Zealand as well as Scotland. He can't remember a more uncertain time in his thirty years of farming. As we unload them he confides his fears. 'Will Europe continue to take our lamb? Perhaps they will but if there are tariffs it could be at much lower prices for us. Can we afford to keep going?' He shrugs[23].

The sheep is perhaps the animal most intertwined with our island story. Yes, I know, the unassuming woolly creatures you count in your head to get to sleep, the muddle-headed animals that stand as metaphors for passivity and indecisiveness. That fundamental component of the British constitution, the Woolsack in the House of Lords, was placed there by Edward III to emphasise the importance of sheep to the British economy. Sheep were behind the controversial Enclosure Acts that were the catalyst for the creation of the iconic Beatrix Potter landscape we know today. And no doubt the monastic flocks played a part in Henry VIII's calculations. Later sheep took the blame for the Highland Clearances as lairds moved crofters off the land to make way for flocks that would produce the mutton to feed Victorian Britain.

---

23. In the event a last-minute trade deal with the EU averted this disaster.

We still have one of the largest sheep populations in the world.[24] No deal from the EU would be a very bad deal for the sheep men as under WTO rules sheep meat has a particularly high tariff of over 40 per cent, outweighing any beneficial impact from a fall in the value of sterling. Although there is increased demand from China, currently being exploited by the nimble New Zealanders, it could take several years to access new markets. Farmers' biggest concern of all is that a move away from protectionism towards free trade will repeat the terrible blow suffered by the rural economy following the repeal of the Corn Laws, and that farmers' livelihoods will be sacrificed in a wave of trade deals with the New World. Our ability to compete is compromised by some of the highest labour costs in the world and a regulatory burden we have inherited from the EU and will probably keep. For example, every sheep is tagged with an individual number, something that not even New Zealand has.

In fairness to the Brexiteers, there was a plan to mitigate the effects of free trade. In the run-up to the referendum the former DEFRA Secretary Owen Paterson presented a paper on post-Brexit agricultural policy that recognised the importance of livestock farming in marginal areas for underpinning the tourist industry and maintaining rural communities. He advocated following the Swiss model that would see generous subsidies continuing for certain sectors, which could be allowed within WTO rules. But will the plan survive contact with the withdrawal agreement? Or the arguments over public spending following no deal?

Meanwhile, there is a revolution going on in the hills and glens of upland Britain. Farmers are giving up their leases and shepherds are being laid off and their flocks culled to make way for trees, as landowners read the writing on the wall and opt for forestry instead, timber is one commodity that is definitely on the up as industries replace plastic with wood fibres and timber is in demand to fuel biomass plants. It is being called the new Highland Clearances in the house journal of Scottish agriculture, the *Scottish Farmer*. This trend is being quietly aided and abetted by both Westminster and

---

24. Of the sheep meat from our national flock of around 35 million (as opposed to only 5.25 million in the USA, who knew?), 40 per cent is exported, of which 96 per cent goes to the EU, mostly to France.

Holyrood. The Scottish government has just cut by 20 per cent the Less Favoured Area Payment, a subsidy which most Scottish sheep farmers receive, in a move widely seen as an attempt to push marginal farming businesses out. The politicians want trees instead.

It is a bewildering time for sheep farmers. They have grown up believing that they are responsible stewards of the countryside, helping to feed and clothe the nation. But, despite politicians talking green, wool is barely worth the cost of shearing as the world persists in using polluting man-made fibres for clothing and insulation: to the point where breeds like the Wiltshire Horn that shed their wool naturally have become popular. And livestock farmers now receive death threats from militant vegans on social media and their animals have to run the gauntlet of demonstrators outside abattoirs.

Part of the problem is that, from a purist point of view, sheep do not belong in the British landscape. Although sheep have been here since God was a boy, they were not here when God was an *infant*. The absolutist Neo-Roo narrative is that sheep originate in Mesopotamia, which is true, and do nothing but damage to our environment, which is only partially true. They rub salt into the wounds of the sheep industry by pointing out how little lamb actually contributes to GDP, and suggesting that a re-wilded landscape without sheep would contribute more through eco-tourism. This argument is possibly more of a reflection of all time low incomes caused in part by meat processors and supermarkets driving farm gate prices down. And certainly if society valued wool more, gave up wearing synthetic fibres and used wool as insulation again on buildings like the Grenfell Tower, instead of flammable plastics, the economics would be very different. It is also dubious to lend too much weight to the absence of sheep in the Holocene era. In Ken Thompson's book *Where Do Camels Belong: Why Invasive Species Aren't All Bad* it is argued that the earth once only had one super-continent, Pangea, and flora and fauna have been evolving and migrating ever since. It transpires that camels really belong in Canada. And many species that were not present in the Holocene period *were* here in previous interglacial periods. And sheep have been here so long that they are ingrained in our culture and have adapted to our environment. The seaweed-eating sheep of North Ronaldshay would struggle if they were sent back to Mesopotamia. It is certainly true that ovines do not contribute to the ecosystem in the way that cattle and pigs do. However, in my travels in search of wilding projects from Sutherland to Sussex and Devon to Norfolk, I came across a number of conservationists who were using sheep successfully as part of the mix. Jake Fiennes, director of the Holkham Marsh National Nature Reserve, believes in having sheep at the end of the grazing season to ensure that the grass goes into winter in the right condition for wintering geese and subsequently for ground-nesting birds in the spring. Sheep are important for producing the short cropped 'lawns' on moorland that many wading birds need for nesting habitat. On my own farm a natterjack toad breeding programme requires me to keep sheep close to their ponds to maintain a tight sward.

The writer James Rebanks, who is probably the best-known sheep farmer in the country after the success of his bestselling book *The Shepherd's Life,* has proved that partial wilding and sheep farming can go hand in hand on his Cumbrian hill farm. He works closely with a botanist to maximise the wild flowers on his pasture and hay meadows, which are grazed by Herdwick sheep and Belted Galloway cattle on a seventy- to ninety-day rotation. They discovered that over thirty plant species had been lost through *lack of grazing*, rather than overgrazing, and set about replacing them. And he makes room for nature by creating small ponds and areas of willow and thorn scrub that he describes as 'bits of Knepp'. He is successfully creating a wildlife-rich patchwork. Rebanks' mantra is that nowhere on his farm should you be more than 300 feet from another habitat, and he thinks of his fields as 'woodland clearings'. No one could possibly describe it as a 'sheep-wrecked landscape'. The sheep debate is just one example of the numerous 'green on green' arguments that surround wilding.

The case against sheep, put forward by eco-activists like George Monbiot, is a complex one rather like the fragile ecology of the uplands. Some of it is Neo-Rousseauism. However, there is some justification in the charge that overgrazing by sheep in the uplands has degraded the environment *in some places*, although this argument is often re-fighting the battles of a generation ago when the EU, in its wisdom, decided to link subsidies directly to sheep numbers and there was mass overgrazing, which caused many hills to become wildlife deserts of bracken and molinia. That damage has been stopped in most areas, although Wales is a notable exception where landowners are powerless to prevent overgrazing by farmers with commoners' rights. But equally it could be argued that under-grazing, or increasingly no grazing, has allowed moorland to revert to bracken and woody heather with loss of habitat for birds. Where there is muck there may not be brass but there are insects and invertebrates, and therefore wildlife. Sheep also comb moorland for ticks, which are then destroyed when they are dipped, improving the health not just of hiking humans but also of birds and wild mammals. The lack of management for sheep (and grouse, which is what really annoys the leftists) also leaves moorlands vulnerable to fires, as happened on Saddleworth Moor in 2018. Monbiot claimed, rather implausibly, that fires are more likely where moorland is managed for grouse shooting, but the scientific consensus was that lack of active management, by controlled burning to produce short heather for sheep and grouse, had led to Saddleworth, which is managed by the RSPB, becoming a tinderbox of rank vegetation.

The implications of a collapse in the sheep industry are far-reaching. Depopulation of remote rural communities may be one consequence as the rural economy shrinks. Monbiot has criticised agricultural subsidies for contributing to a 'sheep-wrecked' landscape but the National Farmers' Union argues that for every pound of subsidy given to British farmers, £5.30 is spent by agriculture in the rural economy. And then there are the bed and breakfast businesses that rely on sheep to keep the land looking green and pleasant for the tourists. Sheep have always been the affordable bottom rung of the ladder for new entrants wanting to farm. The traditional route into agriculture has been for young farmers to buy a few ewes and rent some grazing to build up the stock and capital required to take on a farm tenancy. If they can no longer make it work it will have a deep and lasting effect on the human resources of all British farming.

CHAPTER 5

London is Another Planet

We are in London as finalists in the Eviivo Bed & Breakfast of the Year Awards – I say *we* but really Sheri does all the hard graft, although I am happy to share in the glory. We don't win but it is a happy night swapping anecdotes with other breakfast chefs from all corners of the UK.

London is now culturally another planet as far as this backwoodsman is concerned. There is a gang of misanthropes out on the streets calling themselves Extinction Rebellion. I should have thought if anyone had the right to rebel about extinctions it would be farmers driven out of business by badger-borne TB, keepers and shepherds being laid off in the hills as they are planted with forestry, and keen young huntsmen robbed of their futures by the Axis of Spite.

Having what socialists call a vested interest in the survival of British livestock farming I am anxious to see the vegan revolution at first hand. George Monbiot, the Citizen Smith of our time, has just blockaded Smithfield Market and announced that British farmers are all going to give up producing meat and have a wonderful future growing vegetables. I don't suppose George ever ventures out of his ivory tower in Oxford but if he did he would discover that there are even parts of the Cotswolds where that isn't possible.

I find them in Trafalgar Square. It brings back happy memories of the Countryside March, when our rumbustious throng of rustic dissidents entered the square a blue grey flock lifted off Nelson's Column and some wag shouted 'PIGEON!' and we all looked skyward and shared a cultural moment. I don't know what Nelson would have made of the dreary bunch of hippies sitting on the steps outside the National Gallery with their placards, all seventy of them, watched by seventy-two bored-looking policemen, who have been bussed in from Kent at our expense. What fascinates me is seeing how they are being manipulated by a man controlling a film crew. He keeps moving them so that the camera makes it look as if the whole square is full of protesters. I seem to recognise him and then I remember, he is a left-wing university lecturer, who has just appeared on BBC *Question Time*, where he was highly articulate. I had marked him down as a classic Marxist-cum-Neo-Roo.

They all look very anaemic, as might be expected from a vegan diet. Apart from an eccentric-looking fellow with a banner reading, 'When they circumcised Trump they threw away the good bit.' Obviously a Democrat then. A sad-looking woman cowers behind a large placard that reads 'This feels like a dictatorship.' I look around the square at happy tourists of every creed and colour, at children climbing on the lions and at the random collection of 'environmentalists' being humoured by the avuncular arm of the law and ask 'Really? Do you really think so?' It is always a mistake to get into conversation with single-issue fanatics. Soon I am being proselytised by a woman clad from head to foot in ocean-polluting, synthetic fibres, who assaults me with falsehoods. 'Do you know, eating four beef burgers is the same as flying to New York and back! FOUR BURGERS!'

Always up for a debate, I explain that I am a beef farmer and her assertion is what you might call a double whopper with fries. My grass-fed beef does not harm the planet, and how should the farmers of north and west Britain provide the nation's protein in areas not famed for their lentil production?

'Ah,' she says, 'you should just grow trees.' Folding her arms to reinforce the *non sequitur* in smug, vegan self-righteousness.

I head back to Galloway a sadder, not much wiser man. Our world is under threat. If rising sea levels don't get us then the vegan thought police will. Happily, if Extinction Rebellion has achieved anything, it has galvanised the scientific community into looking more closely at the myths surrounding methane from cows and exposing the fake science. I find a series of seven short 'cows and climate' videos on YouTube produced by GHGGuru, aka Dr Frank Mitloehner of California. I wish I had done so before my trip to London so that the facts were at my disposal.

* * *

Two steps forward, one step back. The bank has turned down our request for a loan for the dairy. It's a bitter blow but Richard and Brendan and I pare the budget down and resubmit it. John is hopeful that we might be okay. It's an anxious wait.

* * *

A book, *On Eating Meat*, comes in for review and I wish that I had read it before tangling with the vegans in Trafalgar Square. Greta Thunberg describes climate change as 'a black and white issue' but to most people it is baffling shades of grey, and nothing is more nuanced than the argument over what we should eat. Matthew Evans thinks and cares very deeply about where our meat comes from, 'dabbled in vegetarianism', trained as a chef before becoming a food writer and then a farmer in his native Tasmania, 'drawn to the land by a love of ingredients, a love of flavour'.

Evans's thesis is that 'ethical omnivores' should care more about the provenance of their meat and he dissects industrial food production and the cruelty to animals 'done in our name'. Since Australia axed farm subsidies and embraced free trade its agriculture has taken shortcuts that would be illegal here and he shines a merciless light on caged chickens grown in the dark then killed at a rate of 160,000 per day in one abattoir with a conveyor belt 4.2 km long. And pork systems where the sows are killed and the piglets cut out of their dead bodies to avoid infection.

But Evans remains an omnivore and he challenges vegans' 'saviour complex' preconceptions. He makes the point that every type of food has an impact on animals, particularly annual monocultures: maybe twenty-five times as many sentient beings die to produce a kilo of protein from wheat than a kilo of protein from beef. And he explodes many of the myths about carbon emissions and water usage in beef production. He is sceptical about the new fake meats, for which, 'there will be a place … just as there's a place for frozen pizza and tinned spaghetti.'

November 2019
I can't do anything about the weather but I still feel thoroughly ashamed looking at where the burn flows onto the beach. It is the colour of chocolate from all the nutrient-rich soil that has come off our fields. It is even more embarrassing up on the potato fields, which now look like a bad day on the Somme. We managed to get all the spuds off thanks to our contractors working through the night when the chips were down, if you will excuse the pun. But sowing the following winter wheat crop is impossible and we have had to concede defeat to Mother Nature. In the meantime, for want of a green cover, the cold, wet soils exhale greenhouse gases and our hard-earned fertility keeps washing off the wounded land and into the watercourses. And that, my Green friends, is the downside of growing 'plant-based food'. Our cattle may belch methane but their grass fields glow with a smug shade of green as they lock up carbon and water under a thick mat of herbage aided by all the organic matter the cows have been spraying on it all year.

It could be worse. My heart goes out to those farmers in South Yorkshire whose potatoes are still rotting in the ground. Potatoes are ruined after twenty-four hours under water. At current prices they are worth nearly £3,000 per acre. I have heard of one grower who still has 600 acres to harvest. That is a big loss for a small business to bear.

Still, at least my soil is not now in some old lady's kitchen. As a budding journalist I confess to a prurient interest in a very British rite, lent extra spice this year by the floods being in marginal constituencies during an election campaign. Cue ministers in suspiciously new-looking wellies trying to look simultaneously sympathetic and decisive; and opportunist opposition MPs trying to see how many times they can fit the words 'national emergency' into soundbites. All we need is George Monbiot to reheat a *Guardian* piece blaming grouse shooting for the floods, and a woolly looking chap saying that they could have been prevented by beavers, and the ritual will be complete.

Floods are toxic for politicians. David Cameron once told me that he had sacked a minister partly because he had 'had a bad flood'. Ministers blame the 'nature of climate change' for the scale of damage. But they are on shaky ground when they say that floods are becoming more extreme. 'The Welland and Deepings Internal Drainage board in Lincolnshire have been keeping rainfall records since 1829', our friend Nicholas Watts[25] writes to tell me, 'And when I look back on the rainfall over the past 190 years we are getting *less* extreme events than we used to. The worst year was 1880.' This tallies with the conclusions of a number of other farmers around the country who keep rain gauges. They know a thing or two about drainage in the Fens, where mercifully there are still some local drainage boards managing catchment areas successfully. 'Every year the watercourses are cleared and every ten or fifteen years the mud is taken from the bed ensuring that the drain will convey the water it was designed to.' In stark contrast the Rivers Welland and Witham are controlled by the Environment Agency and flood.

Every time the same arguments: the towns downstream blame the farmers upstream for allowing the water to run off too quickly. And the farmers upstream blame the downstream agencies responsible for the rivers for not dredging enough to allow the water away from their fields. Our hapless politicians throw money at the problem and hope that the waters will recede before they are sacked. We thought we had sorted this problem when Owen Paterson gripped the Environment Agency and had the Somerset Levels dredged but clearly we haven't learnt the lessons.

The civil servants are simply incapable of understanding that, like it or not, most of our watercourses are now in some way artificial and in need of human action. Their refusal to countenance dredging has resulted in millions of small mammals, insects and invertebrates being drowned with a consequent impact on birdlife, such as the barn owls that prey on them.

It was probably a desire to keep this toxic problem at arm's length from ministers that persuaded John Major's government to set up the Environment Agency and give it authority over the

---

25. One of our B&B regulars and a very knowledgeable farmer and naturalist.

rivers. The trouble is that flood prevention and caring for riparian environments are two activities that are completely inimical to each other. The first is an industrial process involving diggers gouging out sensitive gravel beds and water margins to make the country's rivers carry more water, and the second is a more aesthetic pastime involving leaving rivers undisturbed to create habitats for rare flora and fauna. The second is infinitely sexier as far as the civil servants are concerned and wins every time. And so we have arrived at a situation where the officials responsible for drainage don't really believe in it. A bit like having the Chief Rabbi in charge of bacon production.

The answer is surely to take responsibility for flood prevention, which is a civil defence issue, away from the EA and keep it as an operational function of government, preferably under a Royal Engineers general. The responsibility for river management should be removed from the Environment Agency and given back to local boards. That would facilitate more dredging downstream. The saving on pen-pushers should pay for more sandbags and dredger hours. And we need to focus on our soils. Part of the problem is three generations of reliance on petrochemical fertilisers. We need to go back to using more muck to increase organic matter. So stop burning straw in the Drax Power Station and plough it into the soil instead and, whisper it, more livestock to help keep the uplands like a giant sponge. And, yes, beavers would be very helpful in upland streams (but disastrous further downstream).

And some contrition would be in order.

* * * *

'The pneumonia seems to be just one lug doon this year.'

I nod sagely, I have long since learnt to defer to Davie's superior knowledge of bovine health matters. Davie has a deep disdain for 'books', so when a learned veterinary paper recently announced what it thought was a ground-breaking discovery: that there is a link between an animal's health and how it carries its ears, he didn't see it. But he didn't need to. The research was centuries behind the generations of stockmen who have passed on animal lore like this through an oral tradition that may go back twelve millennia to when we first started keeping animals. A friend of mine once opined that it is more important that a stockman can spot a case of pneumonia from a hundred yards away than that he can spell it.

Scanning the pen carefully it takes me a while to see the animal with its head slightly lower than the rest. There is the faintest tell-tale dark sweat stain on its red flank, which is rising and falling a fraction more than usual. Sure enough it has one ear down. Davie goes on to tell me that his brother, a stockman down the road, has noticed the same with his calves, yet last year they had had both ears down – sometimes one, sometimes the other, sometimes both but always the same. Mere coincidence or the outward signs of the myriad different diseases that trigger pneumonia? Finding out what causes it has been a detective story. Each time we think we have pinpointed the strain, so that we can vaccinate and avoid costly antibiotics, it mutates slightly.

If not treated, the calf may die and, even if it survives, its lungs may be scarred so that it is always sickly. Davie's eagle eyes and stock sense have saved hundreds of lives by allowing early treatment. If April is the cruellest month, when the knacker man is busiest and calving often brings death amid the joy of birth, then November is the one stockmen fear most. When the dank air hangs heavy across the farm and lines the calves' lungs with water, and eternal vigilance for signs of pneumonia can mean the difference between profit and loss, or, in the current beef market, between a small loss and bankruptcy. The poets can keep the 'Season of mists and mellow fruitfulness' as far as we are concerned; we just want to get through it as quickly as possible with this year's crop of calves intact and hope for a cold, clear winter with enough fresh air blowing through the shed.

We always pray for a dry back end so that we can keep the cattle out but every year, as the ground starts to poach, we shift them to sandier fields and then finally we bow to Mother Nature and bring them in before the fields turn to a quagmire and next year's grass starts to suffer. One by one the neighbours cave in, all except Steve who has the lightest land in the parish, almost pure sand, a curse in a drought but worth its weight in gold for the saving on straw through the winter. He can keep his out all year.

At least modern cattle sheds are designed to be well ventilated. There is an irony in the twenty-first-century obsession with 'factory farming'. It started when agriculture began to achieve economies of scale and the old byres of my grandfather's generation were consigned to James Herriot films. I remember them from my youth. Picturesque they may have been but they were dark, musty death traps for young cattle. And the cause of backache for the men, and it was mostly men in those days, who had to muck them out by hand. As a result, the modern farm worker has an innate disdain for jobs that involve getting out of the tractor cab and using a shovel. The happy confluence of powerful loader tractors, steel-framed buildings that are high enough to drive into, and EU grants to pay for them, swiftly led to cattle being housed for the winter in buildings that look like factories. Hence, in part, the lazy assumption that modern farming has become industrialised. But the public needs to be educated to look beyond the architecture. It was the old steadings that were the real dark, satanic mills. And converting them for human habitation is the best use for them.

At least the winter routine brings us into close proximity with the cattle. Even if they are losing us money, there are few things more satisfying than leaning on the feed barrier watching the cows cudding rhythmically with dreamy eyes, and happy calves fast asleep on the straw with full bellies like fat piglets. We just need to keep watching those lugs.

December 2019

You hear them before you see them: Daphne and the Bassets, no not a rock and roll band but Dumfries and Galloway's niche rabbit control service for the discerning landowner. Connoisseurs of hound music appreciate basset hounds for the quality of their voices and we could discern an exciting, anticipatory tone to their choir practice as they tuned up. When I have come across them unexpectedly I have always been startled by how much noise comes from so little canine mass. It is a bit like passing a concert hall and hearing what one imagines to be a full-scale orchestra practising a Wagnerian overture and going in to find a quartet of pensioners with mouth organs and recorders. And so it was last week when we walked down our neighbour's drive thrilling to the sound of a basset voluntary and came round the corner to find the battered white van with seven couple of the aforementioned hounds in the back.

Daphne herself, resplendent in a blue hunt coat and breeches topped off with a Mary Poppins felt hat of venerable provenance, poured me a glass of port. Her first whip, kennel huntsman, white-van driver and husband, in that order, Robert, came up beaming as he raised his tweed cap and shook my hand. 'Fine scenting morning.' The two of them have nigh on a century of hunt service between them, Daphne latterly as our hunt secretary, and in retirement they have gratefully forsaken the saddle and the physical demands of fox hunting for the cheerfully pedestrian business of rabbit hunting. Their bassets have raised thousands for charity as well as thinning out the county's bunnies. They are assisted by John the ferret man, who lurches towards us with a large wooden box slung over his shoulder, just in case all the rabbits are below ground.

The plan is to draw the neighbours' gardens and then to work our way back towards us, and within minutes of hounds being let out of the van they are on a rabbit. If you could see a basset in human form I imagine it would be something like the Bond baddy Oddjob, a muscular thug with deadly habits. I had always been sceptical of their hunting abilities and suspected that Aesop's fable of the tortoise and the hare might have been prompted by a day with the Thracian basset hounds. But there is no doubting their efficacy in close combat. The next half-hour passes quickly as the barbarian horde butchers its way through the shrubberies laying waste to the local rabbit population with savage, ululating war cries.

'I think that's two brace now. I only count the ones I actually see. Quite a few get eaten before we can verify them,' says Daphne as another one bites the dust. 'Shall we go and see if we can find your rabbits now?'

It's just then that I remember that our lawn is double-booked. Barbara and Arlene[26] are doing a photo shoot. Absolutely charming, they are among our more unusual bed and breakfast guests.

---

26. Not their real names. They have now been back to stay with us several times and become friends.

Barbara is a six-foot-two lorry driver, baritone, size twelve feet. Arlene has a five o'clock shadow and a passing resemblance to David Walliams. The whole woke thing has largely passed us by in Dumfries and Galloway but, though we are too polite to pry, we assume that they put the T into LGBT. They had asked over their full fry-up whether we minded them taking some photographs in the garden, and Barbara had emerged, ready for Arlene to snap her, wearing an antique Victorian crinoline with a fifteen-foot bustle complete with accessories including a fan to cool the December air. They are there now as I approach them with all the diplomacy I can muster. 'All right if we bring the hounds through? Just doing a spot of rabbiting.' They appear enthusiastic about the idea and so, as we lay hounds onto the rabbits' favourite haunt under the fuchsia hedge, I am able to reflect on the panto season tableau of a middle-aged woman dressed as Little Boy Blue blowing on her horn to exhort a pack of dwarf hounds while a man dressed as Bo Peep looks on. I had forgotten the inquisitive nature of bassets. The dress looks too interesting not to investigate … and … anoint. If there is ever a time for moral cowardice this is surely it but Daphne, good churchgoing Christian that she is, fesses up straight away.

'I'm awfully sorry. I think one of my hounds has just peed on your dress.'

'Oh that's all right. It's the countryside isn't it?'

Thank heavens that British tolerance and good manners still exist out here in the sticks. Would the same spirit have prevailed if the encounter had been in a London park? Let's hope it still would.

CHAPTER 6

Trees and Geese

The Brexit election is fast morphing into the green election as politicians woo floating voters, particularly millennials, concerned about the environment. Trees are fast becoming the answer to everything. Worried about floods? Don't bother dredging rivers, just plant forests on farmland upstream. Flight-shamed? Get your PR to tweet all the trees you have planted on your Hampshire estate. Want to spend Christmas at Chequers? Make sure the manifesto commits to more trees, and not just the magic money variety. The higher environmental concerns rise up the polling data the earlier 'zero-carbon' commitments become and the more politicians fall over themselves to burnish their tree-planting credentials. I like trees, who doesn't? There are worse things for politicians to blow our money on. They could be promising more traffic wardens or tax inspectors or a state-controlled internet. What's that? Oh yes, sorry, actually Corbyn has promised that one. Much better to dream of bosky woodland glades sucking up all that nasty carbon and saving the planet. Still, I can't help feeling a nagging unease about our politicians' new-found enthusiasm for silviculture.

Jeremy Corbyn has put all the other parties' commitments in the shade with a sun blocking two billion, yes, you heard that right, two billion trees. Never mind that these new trees would need an area half the size of Wales to be planted at commercial densities or the whole of Wales if they are to be grown more naturally in line with his other commitment to 'wilding,' thus radically altering the nature of the British countryside and rural communities. If Labour get in we are all going to be planting trees or facing the consequences from the sinister-sounding new 'environmental tribunals'.

Public opinion is buying the environmentalist agenda and starting to chant the mantra of 'tree roots good, four legs bad' egged on by the Neo-Roos, who have long wanted to turn private grouse moors into public forests complete with lynxes, pine martens and beavers. And the chance to line up veganism, animal rights and 'wilding' like three lemons on a fruit machine must seem just too good to be true to them.

Actually it is all a bit too good to be true. There are a series of lazy assumptions bedevilling policy making. That very British combination of self-importance and self-denial in our national character overstates the effect that we have globally and forces us to sacrifice ourselves. How can we be sure when we plant a tree on grazing land in this country that we are not just causing a better tree to be cut down in the Amazon basin to replace the protein in our diets? Swapping bio-diverse primary rainforest for dark, birdless, conifer plantations on Scottish hillsides is a crap deal for the planet. The 'greenwash' of airlines trading carbon credits with forestry companies smacks of medieval sinners buying indulgences from the Church. How do we know that the trees being planted under these schemes wouldn't have been planted anyway? And do they really make flights 'carbon neutral'?

Counter-Enlightenment 'climate emergency' superstition has trumped scientific rigour and empiricism. Much of it seems to be of the four burgers variety, designed to see how many whoppers the politicians will swallow. The undoubted benefits of trees are often exaggerated and seldom weighed against the carbon costs of ploughing before planting or new roads to extract the timber. Nor are they netted against what was there before, which might already have been very efficacious carbon sinks – either efficiently grazed permanent pasture or moorland that has been burnt regularly to renew the sphagnum moss and heather. The trees versus farmland argument has recently been thrown into doubt by new science from the University of Helsinki showing that pines, birches and spruces actually increase nitrous oxide in the atmosphere.

We think we are returning the land to primordial forest but most academics now agree that it certainly wasn't the closed-canopy forest that some appear to think it was. During the Holocene period, at the end of the last ice age, when the glaciers receded it left a landscape that was probably open savannah dotted with oak and thorn scrub in some parts of Britain and temperate rainforest in others with a heathery montane moorland on high ground above the treeline.

Here in Galloway we have been on the receiving end of successive governments' tree-anxiety since those strategic geniuses in Whitehall decreed in 1919 that there would always be a need for pit-props and timber for trench construction and founded the Forestry Commission. The afforestation of Britain has without doubt led to biodiversity loss. And the dead hand of the state is largely to blame. As both a major timber producer and, unusually, also as regulator of all forestry, including that in the private sector, the Forestry Commission has been responsible for the planting of thousands of acres of conifer plantations on British hillsides. Every time new forestry is planted on our heather hills we swap uniquely British, biodiverse habitat for iconic birds like lapwings and hen harriers – and yes, red and black grouse – for monocultures of alien species like the sitka spruce from the North Pacific islands, which has no wildlife value, even if planting schemes do now require token margins of native trees. Our most at-risk bird, the curlew, will not nest within a kilometre of the forest edge for fear of predators, so it is being further endangered. The overseas tourists who come to marvel at our unique upland landscapes, 75 per cent of the world's moorlands, must think we are bonkers. Maybe we would be better to celebrate what we already have and look at ways to improve our moorland and grassland's already impressive capacity to lock up carbon and retain water, partly by allowing the regeneration of native trees like Scots pine, rowan, birch and willow in an open, grazed, heathery landscape.

The failure of David Cameron's coalition government to privatise the Forestry Commission was hailed at the time as a victory for the environment, particularly by Neo-Roos. But had it been successful it is quite likely that private sector buyers would have been quicker to make the nation's forests more wildlife friendly, and even to re-wild large tracts of them.

The new wave of commercial conifer planting, encouraged by Nicola 'We planted twenty-two million last year' Sturgeon, is upsetting the natives, particularly the consequent 'clearances', a word loaded with folk-trauma in Scotland as landowners reluctantly do the maths, assess agriculture post-Brexit and, with heavy hearts, sell to the forestry companies. The estate agents are valuing planting land in the hills at up to £3,000 per acre, double what it was two years ago.

* * *

The sun starts to disappear behind Criffel and as the poet Gray wrote, 'Leaves the world to darkness and to me'. There are few more peaceful ways to spend an evening than waiting for a duck in the gloaming of a Galloway day with only the owls and the whisper of the wind for company. It is nearly fifty years since the death of my uncle Archie Blackett, in February 1970 aged thirty-seven. Had he lived, in all probability I would not be sitting by what was once his flight pond.

Archie lived and breathed ducks and geese. Living here by the Solway shore it is hard not to be aware of them and he spent every available moment studying them, painting them and shooting them. Just how much time he spent is carefully annotated in his game book, which I now use. Specially made for him in 1960 by Webster & Co., 44 Dover St., W1, it has columns for no fewer than seven species of geese and thirteen species of duck as well as the usual game birds.

He was a good shot, it was his habit to keep a tally of the number of snipe he shot; the last one not long before his death was his 488th, not bad for someone still in his thirties! One entry describes a particularly good drive at Whitfield in which he shot fifty-four grouse. Most seasons saw him out shooting on over sixty days between August and February, mostly rough shooting and particularly flighting. He fed four flight ponds and each flight is meticulously recorded with the time the ducks came and the direction of the wind.

But it was punt gunning that was his passion. My grandfather was very keen and passed his enthusiasm and knowledge onto both his sons. Archie's first shot is recorded as a red-letter day in my grandfather's game book when in 1949, aged sixteen, home from Eton for the Christmas holidays, the two of them went out and shot fifty-four duck in a single shot.

During his time in the army Archie kept a punt on the River Nene and punted in the Wash. But it was when he retired from the army and returned home to farm here that he was able to focus on fowling. Most of his punt gunning was close to home on the Nith Estuary or the Esk, although there are records of occasional forays to Wigtown Bay and the Beauly Firth. He averaged fourteen ducks or geese for every outing and his record shot was ninety-four. Geese were rather easier to sell then than they are now. One of his outlets was the Cairndale Hotel in Dumfries. He would sell them at the kitchen door to the head waiter. After a while he expressed some surprise at the level of demand, given that goose did not appear to be on the menu. It transpired that some of the guests were in the habit of staying in the hotel with their secretaries on the pretext of a wildfowling holiday and they needed some geese to take home to their wives!

Alas, after my uncle's death, the male members of my family bowed to my grandmother's wishes and gave up this most dangerous of sports and my father sold his punt. So my knowledge is restricted to memories of his reminiscences. He would talk dreamily of punting, the numbing cold and aching discomfort, the patience required for waiting for the right conditions, the excitement of spying, then stalking a large bunch of duck or geese, then BOOM, the climax as the gun went off and the punt shot backwards. He would explain that though very occasionally large numbers of ducks would be killed with one shot, ethically it was no different from a day's shooting pheasants with a large number of shots.

They were hardy folk in those days before the advent of Gore-Tex and neoprene. In the bad winter of 1963 Archie's game book laconically records the risk to the punt from icebergs on the Solway.

It was the era of 'gentlemen and players'; amateur enthusiasts learnt from the professional fowlers who earned their living by harvesting geese and ducks in littoral waters around the country and together they formed a unique brotherhood with its own customs and courtesies. One report describes how Archie came across visiting punt gunners on the Solway and despite having 'the right of the river' according to the etiquette of the sport, gave the visitors the first chance at a party of duck that had pitched in downstream. On one occasion Archie was marooned on a large sandbank on the Wash waiting for the tide to come back in when from the other side he heard 'Howzat?' It turned out to be the actor James Robertson Justice with a companion playing cricket using a paddle as a bat.

Sir Peter Scott was another keen punt gunner whom my uncle helped to set up the Wildfowl and Wetlands Trust on the Duke of Norfolk's estate at Caerlaverock just across the Nith Estuary. It is generally forgotten that Scott, one of the great conservationists of the last century – he was knighted for services to birds – was a keen shooting man. He came to the Solway for lessons in the craft on the Cree Estuary. Though a generation apart in age both men shared a love of birds, wildfowling and painting and became friends. Scott did a drawing of geese especially for the front of Archie's game book.

Scott's influence can be seen in Archie's paintings. He was as talented with a paintbrush as he was with rod or gun. Anyone who has ever been out on the Solway can't fail to be captivated by the atmosphere, the freshness of the light and the love and detailed observation of waterfowl he brought to his art. That he is not better known as a wildlife artist must be due to his dying before he reached his artistic prime and the fact that, although he had started to exhibit at the Tryon Gallery in London, most of his paintings remain in the family's hands.

Tragically his passion for punt gunning was what led to Archie's death. On the last day of the season, 20 February 1970, he and his friend Philip Wilson went out for the last time from Browhouses on the Esk. It is reckoned that they had made a big shot when a freak storm blew up from the Irish Sea and drowned them. At his funeral in God's Acre as his coffin was lowered into the ground, the sky filled with thousands of pinkfeet as skein after skein flew past singing at the tops of their voices. This is not an uncommon sight in the parish of Kirkbean in late winter but the mourners were in no doubt that the geese were saluting one of the wildfowling greats. And a sculpture of a goose sits on his grave to this day.

* * *

The winter routine here includes a game of sheep-or-geese. It involves trying to ensure that our winter-keep lambs eat the grass thus earning me the princely sum of seventy pence per lamb per week rather than the geese, which make me diddly-squat. I expect the usual suspects will send me hate mail for turfing the geese off 'land they no longer own' to make way for their bêtes noires – and not just the black ones. There is a bit of inconsistency in this as I suspect that if we removed all the sheep and re-wilded our farm it would revert to thick scrub, which the geese wouldn't like one little bit. One factor in the extraordinary rise in the goose population is surely that they winter so well off the rich grass with side orders of potatoes and winter cereals grown by British farming plc that they go back to the Arctic Circle to breed in prime condition.

Actually, deep down I have a sneaking sympathy with the environmentalists but please don't tell them that. Winters here in the drizzle would be deeply depressing without the magic of goose music; the dawn chorus and evensong would be barren without skeins of geese going inland with the rising sun at their backs to feed then back out in the gloaming to the mudflats to roost. And seeing their flocks grazing our fields is as calmly satisfying as watching over the woolly backed variety. But needs must and, in any case, the geese are well able to look after themselves. 'BB' understood geese perfectly when he entitled his seminal novel about them *Manka, the Sky Gypsy*. (I expect the modern edition, if there is one, is called *Manka, the Sky Traveller* but that doesn't have quite the same ring to it.) The title captures the essential quality of a goose, which is that it has evolved to go where it pleases and eat what it fancies with highly effective reflexes to defend itself against its main predator, *Homo sapiens*.

There was a convenient theory – as most theories seem to be these days – that the grey geese were hefted to one particular parish for the winter. And therefore if the nasty farmers were too hard on them they would suffer terrible hardship. But one year I helped the Wildfowl and Wetlands Trust trap and fix electronic trackers to some pink-footed geese in one of our fields and we are now able to see the whereabouts of 'our geese' online. The experiment has nailed a number of myths. For one thing, they don't stay in the same family groups but rather chop and change flocks at will. And they treat the whole of the UK as one big farmer's market. One day they can be munching our grass the next they can have hopped over to Lincolnshire for a spot of veg. It makes sense that they would go wherever food is available and keep their flight muscles in good shape for the marathon flight back to their breeding grounds in the spring.

Not so the barnacle goose. The entire population of the Svalbard barnacles now spends the winter in either the parish of Kirkbean on the west bank of the Nith or the parish of Glencaple on the east bank.[27]

And any wildfowler who has visited the Solway is familiar with the cry, 'No! Don't shoot! They're barneys.' It's the wildfowling equivalent of finding that the salmon on the end of the line is a kelt or that the hound music at the far end of a covert is a couple of young hounds rioting on a roe deer. It must be galling if you travelled up to the Solway from the south, booked into the B&B, set the alarm for silly o'clock, trudged through the mud in the dark with all the kit and lain in a freezing ditch for hours, thrilled to the sound of geese coming across the foreshore then ... lowered your gun to avoid breaking the law. At least it is a memorable birdwatching experience but I feel sorry all the same.

The barnacle goose, *Branta leucopsis,* is the magpie of the goose world, partly because it is black and white but also because of the grand larceny it performs every winter on local farms. We have just under 45,000 of the Svalbard barnacles and Islay hosts a similar number of their Greenland cousins. The farmers of the parish have a fierce love-hate relationship with them. On the one hand it is a source of pride that we alone host these beautiful birds and winters would not be the same without them. But, when farming margins

---

27. Old bird books suggest they were more widely spread before the WWT Caerlaverock and RSPB Mersehead reserves were set up.

are so tight, feeding the equivalent of around 7,500 extra sheep across two parishes is a considerable burden. Some fields attract geese more than others and the barneys' most popular pastures are like bowling greens by the time they push off back to Spitzbergen at the beginning of May. They have been known to chase the sheep into a corner. The only solution is to try and ensure that the goose fields are grazed first. So it is very irritating to find that the sheep have found a hole in the hedge and are grazing the next-door stubble. But that's sheep for you.

The grass is most needed in the spring when the cattle are turned out and when the most valuable cut of silage is taken. And 'grass grows grass' as the old saying goes, a reference to the need to have some foliage for photosynthesis to work properly and grow what is our most important crop in this part of the world. And also – if you are of a green persuasion – provide valuable service to the planet's climate by recycling carbon dioxide into oxygen. So the barnacles, lovely as they are, are a cost. So too are the grey geese but, partly because we can shoot them, they are dispersed more widely and don't do so much damage.

The protection of the barnacle goose in 1981 was fully justified with a population below 10,000. Since then their numbers have quadrupled in forty years and livestock farmers are becoming the endangered species. Their global conservation status is green/least concern. Their European conservation status is green/least concern. Their UK conservation status is amber. Funny that.

The barneys soon learnt that they were protected. So, interestingly, have some of the cannier pinkfoot who are often seen mingling in the barnacle flocks. We have tried everything: scarecrows, bangers, streamers, chasing them off on foot, but they just laugh and take off to the other end of the field. Shooting a few of them would make them a bit wilder and help to disperse them more.

In fairness to NatureScot,[28] their civil servants do understand the problem and they are starting to grant licences to shoot barnacles. But this involves lots of paperwork, including a laborious process of proving that you have tried scaring them by other means. And the licences are issued to individual farmers rather than to farms.

---

28. What used to be called Scottish Natural Heritage.

And most farmers don't have spare hours in the day to fill in yet more forms, let alone sit in a goose hide.

It would be far better just to take the barnacle goose off the protected list altogether. It could always go back on again. But when I foolishly suggested that we should think about ending their protected status the civil servants looked at me as if I was insane.

January 2020

January used to be one of my favourite months. My natural indolence is nurtured by a slowdown of work on the farm. The cattle have settled into their winter routine. Rather like children at school, calves have already had whatever was going to give them pneumonia. And their mothers are still a few months away from calving. The arable job has been well and truly put to bed; there is no point fussing about the fields: nothing can be done until the soil dries out anyway. And the apparatchiks of the Green State generally leave us alone as there are no crops to measure. Winter is still a novelty; it's one that will wear increasingly thin but for now we can savour frosty mornings and flocks of birds, easily visible, chattering on bare branches. It is a time when farmers can justifiably skive off for a few days' shooting without being made to feel guilty about it. The pheasants may have been thinned out but they are stronger and wilier, twisting and turning in the air to find a gap in the line. The hunting always used to be better in January as 'travellers' leave their own territories in search of mates and then run like stink back to where they came from.

Then, just as mid-winter is bleakest, when keeping animals alive can be a raw fist fight against the weather, along comes 'Sanctimonuary'. This assault on our livelihoods has extra menace now that we know we are up against not just the usual suspects, the Neo-Roos, but the financial firepower of global agri-business as well. Veganism is the best thing that ever happened to the processed food industry and 'Big Food' has seen the huge margins from turning cheap vegetable oils, sugars and carbohydrates into fake meats and milk. 'Woke consumerism' probably started in university common rooms and student bars but its popularity has been very deliberately fuelled by the multinational conglomerates that control global food supplies. They see it as a way of shifting wealth from small, privately owned farms in the livestock sector to giant corporate food producers on arable prairies and in bio-tech labs.

Poor boobies, the Monbiotists and the Packhamites still bear a grudge against farmers for the Enclosure Acts and think they are furthering the long march of the Left by espousing veganism when actually they are being gamed by the wickedest capitalists of them all. They hate to be told this and it provokes outraged tweets accusing me of deluded conspiracy theories. But you only need to follow the money to see the fingerprints of the globalists all over the promotion of science that suits their agenda, like the controversial EAT-Lancet report on diet.

The purchasing power of these global giants buys far more column inches and radio minutes this January than puny meat industry bodies like the NFU can manage in response, and we will all see recipes that use the ingredients that they want us to buy. The nutritionists who warn against the dangers of giving up meat and dairy will be quietly side-lined. That is made possible by a free press funded by advertising revenues. We should be able to rely on our licence-fee-funded state broadcaster for some balance but its egregious institutional bias was revealed in November's *Panorama* programme, *Meat: A Threat to our Planet?*[29] Presenter Liz Bonnin travelled the globe (in a plane presumably) to showcase the most polluting livestock farms and then visited an agri-tech laboratory in Silicon Valley that grows 'meat' in Petri dishes. The clear message was: give up meat. They reckoned without the truth-seeking powers of social media, the next morning on Twitter we saw a clip that had been cut from the programme. It showed the type of pasture-fed beef system we have in this country that demonstrated the virtuous cycle of methane being reabsorbed by grass and explained how grazing herbivores are actually part of the solution to climate change, not the problem. A spooky editorial decision that has yet to be explained.

If I chose, I could plough up all our grass to grow oats to turn into ersatz milk, for example. That I don't is partly out of concern

---

29. A BBC inquiry later found that there had been a lack of balance in the programme but it was too late, the damage had been done.

for the environment. Put simply, livestock farming increases topsoil and therefore takes carbon out of the atmosphere and puts it into the soil; arable farming does the exact opposite, which is why 'environmentalists' bleat about there being only one hundred harvests left on the planet – without understanding that livestock manure is the critical thing that is missing. We can grow grass here better than anything – or probably anywhere – else. And we can do it without needing to buy from 'Big Pharma' the chemical sprays or artificial fertilisers that create nitrous oxide, a far more damaging greenhouse gas in the long term than methane. Why is it that when cows miraculously turn grass into instantly digestible meat and milk the methane temporarily created is demonised, yet when soya is industrially processed into 'meat' and 'milk' we ignore the permanent greenhouse gases caused in growing it then processing it? Could it be because Big Food wants us to think that way?

CHAPTER 7

Progress

Hallelujah! The bank loans have been agreed in principle subject to a number of caveats but it is full speed ahead now on planning for the dairy. Davie and Graham have both been given the option of staying on and working in the dairy or taking redundancy. Both have opted to do something else but will be staying on in their cottages as tenants. It is a huge relief for everyone that it is all being resolved amicably and I am delighted that they will be remaining part of our community.

***

The unfolding tragedy of forest fires in Australia is being interpreted as an apocalyptic portent of a 'climate emergency', justification for the drastic prescriptions of Extinction Rebellion to 'close down the capitalist system'.

Heart-rending tweets from the fire zone convey the raw anger of farmers burnt out of their homes and their interpretation is strikingly different. Many of them, far from begging everybody to embrace the green agenda, are blaming the greens for exacerbating the fires by meddling in the time-honoured practice of burning off excess vegetation to mitigate wildfires. It is also suspected, though unproved, that some fires may have been caused by arsonists motivated by feeding the climate change narrative.

All over the world farmers are being prevented by 'experts' in government quangos from exercising their initiative and carrying out controlled burning to manage habitats that are prone to catching fire. In a bizarre example of newspeak, often the precautionary principle on climate is cited, yet when the partly RSPB Saddleworth Moor, or Caithness's Flow Country, caught fire recently the fire brigade cited the lack of precautions in allowing moors to become overgrown. And these wildfires massively increased carbon in the atmosphere because they set the underlying peat alight. While the Caithness fire was burning it was estimated that it doubled Scotland's carbon emissions for the six days that it burned.

Upland land managers in England are currently in dispute with DEFRA over plans to restrict controlled burning on deep peat. It started with an RSPB challenge in the European courts based on highly questionable pre-2013 science and EU habitat directives. This led to a voluntary code whereby farmers could only burn to strict criteria such as restoring habitat health or reducing wildfire 'fuel load'. But a Natural England position paper in 2019 would effectively 'nail it down so hard as effectively to stop it' according to the Moorland Association. The quango's preferred method of cutting would actually increase fuel loads. Brexit legislation currently commits us to adopting all EU environmental protections initially.

The science is hotly disputed but recent research has shown that so-called 'cool burning', the controlled burning of heather when conditions allow in winter, can – counter-intuitively – increase carbon sequestration by turning excess vegetation into charcoal and stimulating plant growth by regenerating sphagnum moss and moorland grasses that absorb more CO_2. And breeding bird surveys have consistently pointed to strong correlations between endangered wading bird numbers and moorland that has been burnt to produce a mosaic of habitats. The golden plover, in particular, prefers to nest on recently burnt patches; their eggs are even camouflaged accordingly. And speaker after speaker at the 2019 Wildfires Conference in Cardiff spoke of the need to reduce vegetation in vulnerable areas.

Civil servants have so far ignored this research and remain implacably opposed to heather burning, largely, it would seem, because it is a practice carried out to improve grouse moors. It is a classic example of faux science being used to trump the empirical evidence from centuries of experience of shepherds and gamekeepers. From the North Pennines to New South Wales the refrain is the same: we have had enough of experts; let farmers exercise their judgement and there will be more biodiversity, fewer wildfires and less greenhouse gas in consequence, for the benefit of us all.

***

Veganuary grinds on. It reached its apotheosis on Wednesday night with the airing of Channel Four's documentary *Apocalypse Cow*, preceded by a warm-up act, *Meat the Family*. A whole two hours of prime-time vegan proselytising. In *Apocalypse*, George Monbiot prophesied the end of farming as we know it, which was clearly wishful thinking on his part. In *Meat the Family* an ethically dubious reality TV show had young children raising farm animals as pets and then being given the moral choice of whether or not to eat them. *Apocalypse* sounded deceptively plausible – until you analysed it – but beneath the sonorous platitudes and pseudo-Leninist posters, I couldn't help thinking that it was more about profits than prophets. Channel Four, whose ratings have been plunging lately, is alleged to have signed a seven-figure equity deal with The Meatless Farm Company in return for promoting their products – on top of all their other advertising revenues from Big Vegan. Follow the money.

The programme's thesis was based on several big lies, which were all the more effective for being based on half-truths. Unfortunately 'a lie is halfway round the world before the truth has its boots on' and with documentaries like *Apocalypse Cow* wilfully misrepresenting the facts it will be some time before this one is nailed.

A throwaway remark blamed saturated fats from animals for causing cancer and obesity. Actually it is carbs and sugar that are the real baddies but maybe Big Carb is paying the bills so perhaps it's best not to say so. We then had Monbiot, the starry eyed techno-optimist extolling the virtues of growing electric protein from bacteria and hydrogen in factories 'literally out of thin air'. A serious investigative journalist might have pondered how future humans are going to get their daily intake of iron, zinc and other minerals if their food has had no connection with the soil.

There is something darkly funny about a bourgeois socialist opining that the proletariat should subsist on gloop grown in factories. Poor George. He is becoming like one of those deranged Bond baddies with a power lust for seizing the global food monopoly from a secret laboratory in a Scandinavian fjord while simultaneously bringing down the governments of the Western world. It would all be quite funny if it weren't so deadly serious. Farmer suicides are at an all-time high and there are some very vulnerable families worried sick by the constant drip of poison attacking their livelihoods. Channel Four needs to examine its conscience and work out whether it really wants to be seen as a post-truth propagandist for global food processing corporations. Ofcom needs to consider whether Channel Four is fulfilling its public service broadcasting remit responsibly.

As for my fellow farmers, we can only cheer up. There are only twenty-one days of Veganuary to go. And remember that the predicted dystopia of Frankenstein food being grown in laboratories, is neither consumer friendly nor tested safe. The sunlit uplands are ours and their green pastures will be grazed by sheep and cattle for many years to come. Won't they?

* * *

The hunting world has been chipper since Boris's re-election and the South Durham Hunt's Boxing Day meet in Tony Blair's old constituency, now held by a Tory, was I am told, especially cheery. Across the Pennines, in the land of John Peel, Workington Man is pleased to see the back of Sue Hayman, architect of Corbyn's manifesto commitments to tighten the hunting ban. (Hayman would also have been England's first vegan agriculture minister if Labour had won.)

I have a soft spot for Workington, home of the British Cattle Movement Service. The people who work there are warmly helpful when I lose my cattle passports. Workington Man's back garden contains a menagerie of ferrets, racing pigeons, spaniels, Patterdale terriers and miscellaneous long dogs. As you might expect in the land of John Peel, who hunted the fells inland, he is a keen hunting man and is gleeful when hounds accidentally chase a real fox. In fact one of the reasons he voted for an Old Etonian is that he is on first name terms with quite a few OEs through a shared love of field sports – something that his antithesis, Islington Person, can't get 'their' head around. Workington Man dislikes Woke Person intensely. He wants Boris to get Brexit done and then there is an injustice to right: Blair's hated Hunting Act, an injustice that even Blair later acknowledged was a mistake[30], which is at least more than can be said for the only other European leader to ban hunting, the first well-known veggie, Adolf Hitler.

The countryside awaits indications of what the new parliament will do about hunting. And many will feel betrayed if the answer is nothing, particularly after the canvassing done by hunt supporters for pro-hunting MPs of all parties. Especially in Devon where Boris is seen as one of them. But it is not straightforward, particularly inside Number Ten, where Boris and the First Terrier are instinctively pro-hunting – the PM voted against a ban and as Mayor of London provocatively suggested hunting in the capital to control foxes. The First Girlfriend[31] on the other hand is said to be anti, based on her outspoken attacks on trophy hunting and her closeness to badger charities. It will be a key test of who wears the trousers.

Rural lobby groups have been caught unawares by the landslide and the Countryside Alliance appears to be going through a period of navel gazing. Many hunting folk, especially in areas where a Nelsonian eye is said to be cast, would be happy to let sleeping hounds lie. Others see a repeal as essential to preserve hunting and turn the ratchet back the other way after twenty years when metropolitan leftism has held sway over rural life. If it is to be done there is a window of three years before the next election. But it is not straightforward in parliament either. Assuming there are around thirty 'Blue Fox' Tories such as Dominic Raab, who would vote to keep the ban, the new parliament probably contains a majority of MPs who would vote for a repeal, even before you exclude Scottish MPs who *should* stay out of it. But with no

---

30. In his memoirs.

31. Carrie Symonds, later Mrs Johnson.

manifesto commitment – seen by many as a mistake (hunting has never polled as an issue that changes votes either way) – that would make it very difficult. The best hope may be through replacing the Wildlife and Countryside Act 1981, something that has to be re-written once we are outside EU Directives anyway. Foxes have declined by 40 per cent in rural areas since the ban prevented hunts conserving them; and wounding from other control methods has risen.

The vibes from Number Ten are that the government will duck the issue but attempt to placate the hunting lobby with a commitment to strengthen trespass laws to counter hunt saboteurs. Saboteurs are a growing problem, especially in areas where chief constables are not well disposed towards hunting. Clashes are becoming increasingly violent and it may only be a matter of time before someone is killed. Children on ponies are intimidated by hunt saboteurs masquerading as 'monitors'. These 'activists' often have close links to Momentum – 'sabbing' keeps The Mob in training – which is possibly why it abated during Extinction Rebellion. A few years ago a friend of mine in Yorkshire confronted them on his land and had his skull fractured in front of his ten-year-old daughter. His assailants were never charged. Recently Twitter showed a video of sixty-seven-year-old judge Mark Davies and his wife being thrown to the ground by saboteurs – he alone ended up being prosecuted by Derbyshire Constabulary, though later acquitted.

Police resources are being wasted. Strengthening the law is overdue but has no hope of working while *ersatz* hunting continues and provides the justification for 'monitors'. Only a full repeal of the hunting ban will allow the law to stop the violence.

There is a yawning chasm between metropolitan and rural Britain. The current one-size-fits-all, BBC-*Guardian* consensus is not 'one nation'. We need to address the democratic deficit felt by those in rural communities, whose way of life has been steamrollered by metropolitan public opinion over the last two decades. They feel like second-class citizens – partly because electorally they are until boundary reform rights another injustice. In fact if the countryside had the extra twenty seats that are owed under boundary reform there would be no issue with overturning the ban.

I have a treasured signed first edition of Roger Scruton's *On Hunting*, which the great philosopher sent me as an encouragement. It concludes with the words: *If it is true, as I believe, that the fox is better served by hunting than by any other form of cull, and that all rival practices expose him to far more suffering, then it is not just permissible to hunt, but morally right.* Cameron tried to amend the Hunting Act on a far smaller majority; it remains to be seen whether Johnson has the *cojones*.

* * *

Stepping out of the Farmers Club onto the pavements of Whitehall I immediately realise that *Homo rusticus* has evolved differently in his outer skin from *Homo urbanus*. The soles of my shoes are letting in water, something that never happens at home since *Homo rusticus* wears shoes indoors and wellies outdoors. January now seems to be an annual battle to justify our existence to an increasingly metropolitan public. My grandfather would be astonished to learn that this is how I spend Veganuary. In his day farmers were secure in the knowledge that when the leading politician of the day, Winston Churchill, said, 'There is no finer investment for any community than putting milk into babies,' the nation's schoolchildren didn't play truant to demand the end of the capitalist system and compulsory almond milk (produced by the, er, capitalist system). Now there isn't a single farmer in the country who doesn't find himself debating 'environmental issues' on social media or in the snug of the Dog and Partridge (as he still calls it).

Today it is my turn to defend livestock farming on *The Jeremy Vine Show*. I arrive early at Wogan House to eyeball the producers and make them promise that I will not be stitched up – as had happened to me on a radio interview earlier in the week, in a debate with George Monbiot, who was given not just the last word but the first and most of the middle ones as well. The charming girls on the switchboard smile sweetly and inscrutably when I attempt to persuade them not to put any vegan callers through. Schmoozing done I attempt to do some work on my laptop in a corner of the green room but the Wi-Fi in the nerve centre of the world's premier broadcasting corporation doesn't work. The producer tells me it's never any good, which makes me feel better about our hopeless broadband at home. My opponent from planet vegan is a charming girl young enough to be my daughter. She arrives late and flustered, and without a copy of her book she is promoting, so I feel sorry for her and, with possibly misplaced chivalry, mentally cancel some of my attack lines about the prevalence of chronic flatulence among vegans. Then we are on. I needn't have worried, Jeremy Vine is a charming host and scrupulously professional. We both leave the studio genuinely not knowing where he stands on the issue, which is just as it should be.

Back home the sheep have crawled under a gate and are munching all the wheat that the geese haven't had. I relax by walking on the shore with the dogs. The knot murmurations have been dazzling this winter, more so than the better-known starling displays, as the knots change colour like the leaves of a cricket bat willow when they turn. They are the Solway's winter ballet, performed to the accompaniment of curlews and oyster catchers piping across the estuary. This year they have mastered a new choreography, splitting into two identical clouds of swirling, silvering stars before coming back together again. Suddenly their mood changes and they form an angry flying saucer that repeatedly smacks the surface of the sea, the beat of their wings making an aggressive insect noise before bouncing off like a skimming stone down the tide. Seconds later a seal pops his head up wondering what all the fuss is about.

February 2020

So that's it, we have left the EU. The immediate question is what to do with the signs around the farm? These proclaim the generosity of the European Union for funding various 'schemes' under the familiar blue and gold flag (thereby letting the neighbours know what a subsidy junkie I have been). I consider taking them down and storing them safely so that I can fondly imagine my great-grandchildren taking them on the *Antiques Roadshow* ... Back in the early twenty-first century before laboratory food there were people called farmers who were paid public money for public goods by an empire called the European Union and then along came Brexit and the British government changed the policy to, er, public money for public goods. But my instinctive fear of officialdom prevents me in case some apparatchik of the Green State comes and asks for all the money back.

* * *

At last, a hint of spring in the air. There is a hesitant dawn chorus as if the birds can hardly dare believe they should be singing again; the woodpecker can be heard above the drifts of snowdrops in the woods, and the foxes have one thing on their minds – apart from where their next meal is coming from. A dog and a vixen were so besotted with each other that we saw them trotting along the tideline together at ten o'clock one morning. There are tantalising signs in some fields that the grass is growing but in others the sheep are starting to get 'wandery'. It always happens at this time of year. The logical conclusion is to blame the lack of grass – and by extension the greedy geese – but I think it's fair to say that the angelic lambs that stepped timidly off the back of the float in November have reached the bolshy teenager phase. More time is being wasted putting them back where they belong or extricating them from fences. The latter task involves straddling them and holding them between my knees while I gently ease their heads back through the net as they jerk back and forth vigorously. It is best done early in the morning before there are too many people about to see.

The decision to start selling off the beef herd means that this is the first February for many years that I have not been selling breeding cattle in Castle Douglas. But I was pulled like a moth to a flame to the mart anyway. I love going there for the theatre, and so, judging by the crowd of retired farmers there most mart days, do half the pensioners in the county. It will be a sad day if in future, as seems likely, cattle are all sold over the internet by video as they do already in some countries. Watching a farmer in the ring with his cattle is like seeing a chef anxiously offering up his signature dish for judgement on *Masterchef*. Your performance as a farmer is on display for all to see. The first time I did it I suddenly felt as full of stage fright as my first time in a school play. I became acutely aware that my corduroys were a little too bright for the assembled company and the peak of my tweed cap a little too jaunty. Then there is the challenge of moving cattle around the ring without the ignominy of being kicked or charged. Heifers that were quiet enough at home can suddenly become like wild buffalo under bright lights with all the hullabaloo of the auction. Farmers do it partly to demonstrate how easily handled they are and partly, as my friend Piet says, to 'stop the bugger opening the gate and letting them out of the ring too quickly so that the buyers don't get a good look at them'. It can backfire rather quickly if, as I have often seen happen, the vendor has to retreat behind the safety barrier to avoid being knocked down.

It is always best to avoid selling a single animal for that reason. They are much quieter in pairs – but then there is the challenge of matching them evenly. The slightest difference and buyers will focus on the inferior of the two and bid less. Then sure enough some Smart Alec will come up afterwards and say, 'Aye your heifers were guid but you shouldnae hae put yon wee dumpy yin in wi your best yin.'

The psychology is what we come for. Often it will be the culmination of a year's work and a big chunk of the income on the farm for the year all on one day for the seller, whose wife is often watching anxiously from the seats, waiting to see whether she can book a holiday. Then there are the buyers hunched over the rails round the ring like Mafia godfathers. For them also there is pride and profit at stake. They say you make your margin on an animal on the day you buy it. Their skill as stock judges has been honed since birth through childhood trips to the mart with their fathers then through Young Farmers' Club competitions and college. Their skill at bidding is also on display, or rather not on display. Not for them the fancy numbered paddles of the fine art auction rooms. It is a source of pride that only they and the auctioneer actually know who is bidding. And the auctioneer not only has to have the patter of a second-hand car salesman but be able to know everyone in the mart by sight, the names of their farms and – tricky in Galloway – how to pronounce them.

The seller wears a poker face as the auctioneer starts his recital. 'Big stuffy heifer here. Look at that, a real cow getter. A thousand to start? Eight hundred then? Six hundred surely? Six hundred. Get a bid. Get a bid. Six hundred. Seven. Eight. Eight hundred ...' The rhythm of the auctioneer's mantra is hypnotic as it gathers speed then falters. Sometimes the poker face slips when the bid sticks on a low price. The mental calculation is written on the man's face. Does he withdraw the lot in the hope that he might make a sale outside the ring or risk taking the animal home, paying the haulage and then quarantining at the other end in the hope of a better sale, or sell at a loss? A hint of urgency enters the auctioneer's voice as the seller approaches the box. 'Are you selling today?' The man shakes his head. 'No, not today, sorry.'

At other times the price soars and the crowd's mood starts to buzz as the bids climb. Then the poker face becomes a struggle as he suppresses a smile and I remember my friend Ali's[32] advice, 'Look miserable. It doesn't matter how ridiculous a price you are being offered, you have to look miserable otherwise they will think they are paying over the odds and stop bidding.' Triumph and disaster at the mart, the best free entertainment there is.

* * *

When our children were small I placed a sign on the bend in our drive that reads DEAD SLOW, FERAL CHILDREN. Now that they are adults there have been representations to have the sign removed but we remind them that they are still our children and even more of a hazard now that they can drive. The sign was partly an affirmation of the belief that rural childhoods are supposed to be a bit feral. Although I have a slight sense of parental guilt that our children did not have quite the freedom to run as wild as we did. And a letter I received last week informs me we definitely didn't have the freedom that my father and his siblings had. In fact these days they would doubtless be taken into care. And my grandmother, who is still revered across three counties by all who knew her as a Top Mum, would be prosecuted and sent to parenting classes. Paul

---

32. The late, great Alasdair Houston of Gretna, RIP.

Gunn, now in his late eighties, who lived nearby during the war, takes up the story:

> To keep us out of mischief the Blacketts, Landales and Gunns hired a governess, Miss Nottingham. We really were awful to Notty as we called her. Your Uncle Archie would bring his air rifle to lessons and shoot matchsticks under the table at Notty's legs. David's game was to hide in the bath and when the poor lady went to the loo he would jump up like a jack-in-the-box when she sat down. I recall a time when Archie got his hands on some sticks of gelignite. He had seen it being used to blow out tree stumps on the estate so he knew the modus operandi. We set out and had a happy time with gelignite, percussion caps, fuse and matches and found we could blow quite large trees out of the ground. The only problem was your great-grandmother wanted to know what the noise was. We, of course, knew nothing and promised to find out, hoping that she would forget about it, which, thank goodness, she did. Your grandfather was away at war but I recall a time when he was home on leave. I was looking over our garden wall across the shore at Carsethorn when I saw what I thought was a tree trunk floating down the river. I had my .22 at hand so started to shoot at it. It was a little out of range so I was content to splash shots round to see how close I could get. What I didn't know was that it was in fact your grandfather and a friend out punt gunning and when a tall figure suddenly sprang up waving furiously my heart nearly jumped out of my mouth. I had put your grandfather under fire and was frightened stiff that there would be terrible retribution but to my great relief I never heard a word.

Happy days.

* * *

Until recently you would have put me down as a climate change 'luke-warmer': I live by the sea and I'm very concerned about the threat of rising sea levels, even if I don't buy the more extreme predictions of 'climate emergency' talked up by Marxists and teenagers. But now I am toying with the idea of coming out as a ... 'denier' (sharp intake of breath).

This agnosticism is partly to annoy 'Ice' and other youths who have been trolling me on Twitter. They have been calling me a 'fossil' and a 'boomer' and telling me that I will be 'dead soon anyway leaving a f****d planet for their generation' – apparently it is all my fault. It is also partly because if Extinction Rebellion has achieved anything it has forced us all to analyse the case for anthropogenic climate change carefully. And the results have been boggling. Patrick Moore the former President of Greenpeace is the latest luminary to come out and rubbish the science. He says he 'fears for the end of the Enlightenment.' Amen to that.

Actually I now realise that I have been doing more to save the planet than I had believed. I rang one of the biofuel companies that takes the tallow from Scottish abattoirs to ask about the process. It comes down to this: if I have a bull that dies weighing one tonne, the knacker removes him, the carcase is rendered down and the tallow is turned into biodiesel. How many litres of biodiesel do I get? The answer is 180 litres. Who knew? Vegans are driving cars fuelled by cows!

* * *

'Dad, you are to go and have your cholesterol tested, immediately. That breakfast is soooo unhealthy.'

Our daughter Rosie, always brimming with filial love, is giving me a foretaste of what my dotage might sound like. At least I now have a book I can quote in defence of my preferred breakfast of two sausages, two rashers of bacon, two poached eggs, a grilled tomato, mushrooms and buttery spinach, followed by natural yoghurt with a sprinkling of muesli and some fresh fruit.

The extraordinary global neighbourhood that is Twitter has linked me to Charlie Spedding. He asks me if I will consider reviewing his book *Stop Feeding Us Lies*[33], which he has self-published after being rejected by several agents. It is refreshingly contrarian, a

33. https://stopfeedinguslies.com/

healthy dose of common sense after the blizzard of lifestyle books that the publishing industry has served up recently, full of faux environmentalism and vegan dietary choices. For years I had been dimly aware of lone voices in the wilderness – medics like Tim Noakes, Frédéric Leroy, Paul Saladino, the pioneering ecologist Allan Savory and the greenhouse gas guru Frank Mitloehner – telling us that the received wisdom on diet in the developed world, and by extension the way we produce our food and its effects on the environment, is mostly complete bollocks. Spedding draws together the inconvenient facts and the unfashionable theories and explains with patient logic why we should stop accepting at face value what we are told by governments, mainstream academics and the siren voices of Big Business and start developing a healthy scepticism. Because we have been conned.

When I ring him, Charlie turns out to be a mild-mannered grandfather who immediately strikes me as a grown-up who should be heeded, one with no vested interest beyond a philanthropic desire to seek after truth. He is a retired chemist who has spent a lifetime at the coalface of the health service dispensing advice and drugs to patients in a community pharmacy. And in his youth he was a top athlete. In the golden era of long-distance runners from the North East, Durham based Spedding was up there with Brendan Foster and Steve Cram. In 1984 he won the London Marathon and an Olympic bronze medal. The deduction is that he might just know a tad more about nutrition than the host of writers with PhDs in the subject, particularly as – crucially – he is immune to the groupthink spawned by the peer review system and the funding of research. He says his motivation to take early retirement and write the book was triggered by a recognition that he had been a pawn of Big Pharma, selling drugs to patients who weren't going to get better because they were being advised to eat the wrong things.

His thesis is simple and grounded in anthropology and physiology. We are omnivores who evolved by eating proteins and fats from meat. That is why we have larger brains and smaller guts than other primates. We took a wrong turning when we started making food from seeds – bread and pasta from grain, cooking oils from vegetable seeds like soya and sunflowers. Then we compounded the error when, in the seventies, under the malign influence of Big Food and influenced by dodgy science, Western governments issued nutritional advice that demonised saturated fats from meat, dairy and eggs and promoted carbohydrates as healthy foods. The result has been an obesity epidemic, a dip in fertility and soaring diabetes, cancer, heart disease and dementia.

Most compelling is the graph showing the rise in obesity, the timing of the change in dietary advice coincides exactly with the start of what we now call the obesity crisis. Spedding says we should eat more meat. Husbands everywhere will be able to tuck into bacon and eggs and put lashings of cream and butter on their food and have a counter to their wives' strictures – it turns out that most of us need more fat not less.

The conspiracies behind what historians may come to know as The Gross Dietary Error (my words) are film worthy. Spedding shines a pitiless light on the villains: Ancel Keys the man who persuaded the US government to give the bad advice, the Seventh-Day Adventist John Harvey Kellogg who invented the eponymous breakfast cereals to suppress libido, and the food-tech venture capitalists of today who pay for science promoting veganism to 'take down the meat industry'.

Doubtless the health establishment and journalists in the pay of Big Carb will queue up to denounce this book – if they can't get away with studiously ignoring it. Loftily, and very annoyingly, 'experts' will quote reams of 'research' because these days anything can easily be refuted with 'science'. And that in a sense is the issue. We now have a bloated university sector full of academics who will produce whatever false orthodoxy you want for a fee. The Enlightenment values of reason and empiricism are being eclipsed by Counter-Enlightenment superstition and the witch hunting of heretics who dare to dissent. Before endorsing the book I asked a nutritionist friend to vet it. She said that it was what she and many of her colleagues had been saying privately for years but none of them dared say publicly for fear of losing their licences to practise. Enough said.

CHAPTER 8

Covid

March 2020

This spring will be remembered for record rainfall and coronavirus. All we need now is the Four Horsemen of the Apocalypse to come trotting down the drive. It's normally around now we have our annual pre-natal consultation. This consists of my asking anxiously of the cows, 'Anyone bagging up yet?'[34] And Davie growling, 'They'll come when they are ready.' Which is true of course. Although sadly we had a false start when a heifer ejected a dead calf a few weeks ago when neither we nor she nor the unborn foetus were ready. This provoked days of anxiety while we waited to see if this was the start of an abortion storm that would wipe us out. There's always something, and as farmers love to say, 'If you have livestock you have deadstock.' Which is the agricultural version of the ancient military saying, 'If you can't take a joke you shouldn't have joined.'

Speaking of global pandemics, these crises serve to remind one that in the globalised world no man is an island. A butterfly flaps its wings in the Amazon rainforest ... and so on, you get the picture. Except in this case – allegedly – a Chinese chef with a taste for the exotic serves up bat fricassee in Wuhan[35] and our tractor ends

---

34. This refers to the udders extending as they fill up with colostrum in anticipation of birth. All mammals do this as well but I have learnt that it is most unwise to use this expression anthropologically, or even to refer to it at all in female company.

35. This was the initial assumption in the media.

up being grounded. It seems the vital spare widget that goes in its gearbox is stuck somewhere in a container while the supply chain goes into lockdown. This has caused me to look a bit of a chump as I had elected to go down to one tractor when the beef price collapsed and I now have to go cap in hand to neighbours to rent their machinery to see us through. Oh well, perhaps this supply failure will focus the politicians' minds on food security and they will give British farmers a reprieve after all.

Where there are snakes so also are there ladders; we have started seeing an uplift in holiday cottage bookings. A couple rang in a panic saying they had lost their holiday in Italy and did we have a cottage they could rent immediately. It transpired they had arrived in Italy and just had their first ski down the mountain when they were unceremoniously bundled onto a flight and sent back to Cumbria. As there is no need for any human contact, we say yes. Optimists among us postulate that so many metropolitans will be forced out of cities to self-isolate in remote holiday cottages that when the epidemic ends they will all wonder why they ever had offices in cities in the first place and look around for properties to rent on a permanent basis. I expect there will be a snake again soon and either it will be announced that there is a massive coronavirus hotspot just down the road causing cancellations, or the satellite on which our broadband depends will fall out of the sky and the metropolitans will swiftly revise their plans.

* * *

I am reviewing *Daylight Robbery* by Dominic Frisby for *The Critic*. The clue is in the title: daylight robbery was how opposition politicians described the window tax.

Frisby is very interesting on the threats to the nation state caused by globalisation and the rise of trends that will potentially damage the Chancellor's tax take: non-doms, digital nomads, corporations that either deliberately never make a profit or manage their affairs in such a way that the profit just happens to be made where tax rates are lowest, and alternative currencies like Bitcoin. It is a wretched irony that here in Scotland, the nation that produced two of the world's greatest economists, Adam Smith and John

Cowperthwaite, we have higher rates of income tax than the rest of the UK for middle-income earners – and consequently shortages of doctors and other professionals – thanks to a now ex-finance minister who had never heard of a Laffer Curve. The one consolation, though it doesn't say so, is that the answer to the Scottish currency question may have an answer. The SNP is split on whether they would 'keep the poond' in the way that bankrupt Latin American countries use the US dollar, without having any control over it, or whether to start their own currency, presumably with no reserves to back it. If Frisby is right, the answer now seems to be that most people would simply use Bitcoin and leave the new Scottish government powerless to raise any revenue.

* * *

The news is full of food shortages and Sheri comes back from Dumfries without many of the items on her list and starts digging over the vegetable patch. Like all farmers I have a Malthusian fantasy that food shortages will send prices rocketing and we will be heroes of the nation again. There has certainly been enough panic-buying to cause shops to run out locally, but this is probably because food is being bought more quickly than retailers can restock. I have seen no evidence that we will actually run out. But then neither have DEFRA or the British Retail Consortium been able to tell me how much of a stockpile we have. That data is not held anywhere and we cannot stockpile fresh food in any case. However, we can eat down our cheese supplies by maturing cheddar for less time and we can pull commodities like butter out of freezers. It is lucky that companies had been carrying extra inventory because of fears about Brexit but those stocks are finite and there is now a scramble in Whitehall to quantify them.

The likelihood is that the epidemic will dissipate before we run out of food. But it may still give us a fright over the reliability of our food supply. The saga with our tractor has brought home to me just how fragile the global supply chain is.

I am sure that British famers will not allow a virus that for most will be like a dose of the flu stop them producing food. Cows will still be milked and stock cared for and the idea that we and, crucially, our vets won't somehow manage to carry on doesn't enter our heads.

The worry is a bottleneck further down the supply chain, either through sickness, or because many workers in food processing are Eastern European and anecdotal reports are that many have gone home. This is a vulnerability. Then there is the logistics at every link in the chain before you get to the shelf-stackers.

The DEFRA Secretary George Eustice stood next to British Retail Consortium CEO Helen Dickinson at Saturday's press conference and reassured the nation that we would have enough food. DEFRA has made plans, including relaxing competition rules to allow supermarkets to pool resources and emergency measures to ensure the human resources are in place. But Eustice and Dickinson both know that the supply of physical commodities is beyond their control. When I telephone to research an article, both of their organisations admit to me that their confidence is based on an assumption that food will keep flowing from Ireland and the Continent by road. We are 50 to 60 per cent self-sufficient in food and we notionally run out of UK-grown produce on 7 August every year. But that masks a huge reliance on imports for commodities like fresh fruit and vegetables at this time of year. Shipping is severely affected with ships stuck in quarantine all over the world.

One thing we can say is that we have dodged a bullet. Had Brexit happened a few years ago and had we listened to some bright sparks in Whitehall and abandoned protectionism completely we might not have any agriculture at all judging by a leaked email recently from Dr Tim Leunig. The Chancellor's senior economic adviser argued that Britain could become 'like Singapore' and import all our food. Leunig – an economics lecturer at the LSE on loan to the government, who is said to be a powerful voice in Whitehall – claimed that agriculture makes a negligible contribution to the economy, and described the former British colony as 'rich without having its own agricultural sector'. It struck a raw nerve at a time when morale was at a low ebb owing to low prices.

Perhaps we shouldn't get too upset by the utterings of maverick advisers. After all, we won't avoid groupthink in government if we don't have people like Leunig spouting drivel and thereby playing devil's advocate. No one really believes that Singapore wouldn't have its own farming sector if it could. The former colony had a rude lesson in 1942 in how vulnerable it is. But reading between the lines of the leak (who sanctioned it?) it was possible to discern the Whitehall machine pitch-rolling some unpopular policies. Hilary Benn, a former DEFRA Secretary (and, less publicly, the scion of an old Essex landowning family) saw through it straight away and tweeted: 'I'm afraid there has always been a strain of this thinking in the Treasury. It is profoundly mistaken. Agriculture is a strategic industry.'

Benn is right. We should heed the warning we have been given by coronavirus because it demonstrates what could happen in a future war. The threat we faced from U-boats attempting to starve us into submission was only averted by the cracking of the Enigma code. A generation on, technology has changed and it won't be U-boats next time. Instead we might see drone and cyber attacks on the small number of supertankers that we rely on for our food and a dirty bomb in a suitcase causing chaos at each of the major ports. And the very small number of abattoirs and milk processors, a fraction of what we had in the forties, would also be very vulnerable.

We are very lucky that this is happening in the spring when the cows start to produce more milk; there are last year's lambs to kill and British veg will soon be available. We could boost production by a temporary cessation of greening measures – currently farmers are still being told to keep 5 per cent of land fallow for this growing season. The government is desperate to avoid the R-word but food rationing is already in place in all but name. It's just that in 2020 we don't need ration cards because nine supermarkets dominate UK grocery supplies so completely that the job can be outsourced to them. We are already seeing food being prioritised for the elderly and NHS workers who have been allotted special shopping times. If we get through this without some shortages it will be a very close-run thing. Next time a pandemic might hit in the autumn and cause greater shortages.

George Eustice has brought forward a post-Brexit Agriculture Bill that controversially does not classify food production as a public good. Will memories be short or will it be re-drafted with self-sufficiency at its heart?

* * *

There are moments, as a small business owner, when you are like the captain of a ship faced with the decision whether to abandon ship or man the pumps. Coronavirus has caused more of these moments than anything I can remember. Last week it was pub landlords, this week it is the turn of the self-catering sector to decide what to do. We closed the bed and breakfast side of our business weeks ago as that seemed a no-brainer. The decision on self-catering holiday cottages is made much more difficult for us by the lack of binary guidance from the state and the considerable nuances over social distancing in the calculation.

It is becoming a very toxic issue in some parts of the countryside. Just as the wave of panic-buying in the shops has exacerbated food shortages and led to hoarders being vilified, so the crowd reaction of metropolitans wanting to escape to the country has stirred up atavistic tribal emotions from deep in our folk memory for those of us who would be on the receiving end. And in Scotland this has political overtones with nationalists tweeting demands to shut the border at Gretna. As in all previous epidemics, particularly the Great Plague of 1665, there is xenophobia caused by a natural anthropological instinct to keep the infection on the other side of the moat. This may be wishful thinking; we live in a very remote district but we are far more integrated with the outside than in previous generations; we have an airline pilot living locally as well as other weekly commuters and young people returning from university. And I know of several people who are self-isolating with suspected cases already, from which they are recovering. Concerns are also being voiced over the pressure on local services that tourists bring with them. I would guess that half the households in the parish contain seventy-plus-year-olds. Rural communities like ours have hospital services that are spread thinly over wide areas. It must make sense to avoid overloading them, particularly on the islands.

The knee-jerk reaction is to play safe and close everything down. On the other hand the memories of the deep damage done to the rural economy during the foot-and-mouth epidemic in 2001, when it is now generally acknowledged that it was probably a mistake for the politicians to put out a message that the countryside was 'closed,' make us pause and consider. So far we have managed by insisting that guests arrive later and depart sooner to achieve a quarantine period between lets. There has been extra disinfecting by cleaners in masks and all social interaction has been outdoors and at a distance. Guests have kept themselves to themselves and walked in the hills or on the beach. Anyone who is showing any symptoms or is part of a vulnerable group is not coming. This has been consistent with the official advice and I don't believe this has led to any greater risk locally, although it is of course possible that they have spread germs in shops or filling stations that would otherwise not have been there. And it must make sense to keep parts of the economy going if we can. But clearly now we need to review the situation as the situation worsens and public opinion shifts towards wanting a lockdown.

We need decisions. In Saturday's press conference DEFRA Secretary George Eustice ducked the issue when questioned. On Sunday SNP politicians broke ranks and said that people should stay away in personal pleas in interviews that fell short of being official policy. But many self-catering businesses cannot afford to close down without any assurances of a lifeline from the government. It doesn't appear that anyone in Whitehall or Holyrood has given it much thought so far.

* * *

One blessing of the pandemic is that the planning department seems to have streamlined their operation by saying yes to things rather than being difficult. Either that or the government has told them that they must err on the side of positivity if the economy is ever to recover. Whatever the reason, they have agreed our plans for the new dairy and we can start work. Another known unknown ticked off.

* * *

What a difference a couple of weeks makes in the battle of ideas. It was shaping up for the moment when the deep doctrinal divisions underlying Brexit started to show their true colours. Leave versus Remain was really just a cover for the old antagonism between free trade and protectionism that has been rumbling since the arguments over the Corn Laws in the early nineteenth century prior to their repeal in 1846. We were – and still are, once the current emergency has passed – shaping up for the mother of all food fights. It had seemed that the Brexiteers had the upper hand and we were being softened up for trade deals that would see the phasing out of agricultural support and chlorinated chicken on British plates soon. Then, from nowhere, Covid-19.

The truth is that the real effects of Brexit, good or bad, will be felt not at North London dinner parties where they are most discussed but on British farms and in the market towns that support them. Unlike Sir Robert Peel, Boris has kept the Tory Party remarkably intact so far (minus Letwin, Grieve, Soubry and Co., remember them?). The true test will come when MPs have to sell the idea of a future without subsidies or tariffs to their constituents, especially those, like me, who are farmers. Despite an eighty-seat majority the Conservative Party could soon start to fracture over the issue. The free traders' job has just become much harder. Suddenly food security doesn't seem such an anachronistic concept and the public now experiencing empty shelves, many of them for the first time, might conclude that a degree of protectionism wouldn't be a bad thing after all.

I think it is worth recording where we are. British farming has been stripped down to the bones in manpower terms. We are about to start calving. If I and Davie get the virus we will have to keep going, no one else is going to come and calve the cows, which is a very dangerous job that can't just be done by someone made redundant by the hospitality industry. By hook or by crook meat and milk will get to the farm gate because the people involved are either the business owners or employees who are completely devoted to their jobs. After that it depends on workers showing up for work in abattoirs, milk processors and meat cutting plants. And the logistics and the retailers come next.

There is fear and contempt in equal measure for the soft handed PPE graduates wishing to bet farmers' livelihoods on the theory that, like after the repeal of the Corn Laws, historians will garland them with the praise generally heaped on Peel, Cobden and Bright for delivering a similar post-Brexit dividend for the British economy. The answer, I believe, is to consult Adam Smith. The Prime Minister referenced Smith in his landmark Greenwich speech in February, which was under-reported as the BBC was cross with Boris at the time and did not judge that we needed to hear it. But it was one of the most important political speeches of the post-war era in which the Prime Minister set out his vision for Britain as a confident trading nation. It is a vision that I and many other farmers share. We are excited by the prospect of selling cheese to the United States without a 17 per cent tariff and lamb to the emerging markets of Africa and Asia. It's just that we felt there were a number of circles that had to be squared to get there. The repeal of the Corn Laws is hailed as a 'good thing' on PPE courses but it crippled the rural economy right up to the First World War. The Agriculture Bill seems set to repeat the mistakes.

Adam Smith, wise old bird that he was, extolled the universal benefits of free trade in *The Wealth of Nations* and said that governments must adopt it but cited two important exceptions to the rule. The first of these was national security. Smith was talking about shipbuilding for the Royal Navy rather than food security. Before the railways opened up the Great Plains and refrigeration put South American meat on the menu, not even Smith could have envisaged that global markets could provide our food in the way they can now. I believe he would say that we should secure our national food supply first by protecting local production before exploiting overseas markets. That need not be a show-stopper as far as free trade is concerned. WTO rules allow nations to operate some tariffs and quotas to use their own produce first before going into the global marketplace to buy it where it is cheapest. And tariffs and quotas would be the right way to go and so much simpler and cheaper to administer than the present Stalin-lite system of subsidies administered by an army of civil servants in the Green State.

Smith's second exception was where domestic taxes put our own industries at a disadvantage. He felt that it was reasonable to ensure fairness by imposing tariffs on overseas goods to ensure a level playing field. For this read our very high cost of government in this country, social taxes like the minimum wage and the regulations that rightly insist on very high welfare and environmental standards on British farms. Dieter Helm's suggestion of a carbon tax on goods coming into the country that have been produced at greater cost to the environment must be the right way to go.

Adam Smith also had some trenchant things to say about monopolists. Post-war governments have lazily allowed cartels to operate at both ends of the agricultural supply chain, aided and abetted by the EU. The farmer share of the retail price has been allowed to be at barely above subsistence levels. This has stifled investment and put British farming in a bad place from which to build our trading place in the world. That must change and the cartels in agricultural inputs, food processing and retailing must be broken up.

So I am coming round to free trade but on Adam Smith's terms. Let us restore our self-sufficiency in food to around 75 per cent, which was its post-war norm. And let's use the opportunity of Brexit to encourage British farmers to meet the UK's dietary requirements better. That means ensuring that we will have fresh fruit and vegetables by making it possible for farmers in the south and east of the country to grow them. And let's use our climate to its best advantage to grow meat and dairy products from grass in an environmentally sustainable way in the north and the west. British farmers have the soils and the climate to grow food in a carbon neutral way that Californian almond orchards and Texan feedlots cannot.

* * *

Order: Counter-order: Disorder is a sound military maxim that has stayed with me since Sandhurst. And, just as we are starting to build the dairy, that is what we have had this week in the building industry. The government has to strike the right balance between saving lives now and saving the economy, and therefore lives later once the pandemic has passed. We must not have a Pyrrhic victory where more people die in future because the country is bust and cannot fund the NHS properly. It will have been partly that calculation that led the cabinet to decide that construction sites must stay open as far as possible with social distancing being maintained. It will partly also have been the knowledge that the building industry is very fragmented and full of self-employed contractors, and therefore very hard to compensate.

There was unanimity on this and the policy was agreed in all four nations. Then Nicola Sturgeon broke ranks. She appeared to lose her nerve at a press conference earlier this week and announce that building sites must close. Whether this was under pressure from the unions to protect workers or just because she is determined to follow a policy of differentiation is not clear. Later she corrected herself and suggested that the economy had to keep going and now she has changed her mind again and confirmed the policy for Scotland is to close down the construction industry with the exception of hospital building. She may come to rue that exception as it has drawn comparisons between Matt Hancock's 4,000-bed pop-up Nightingale Hospital and her executive's failure to open hospitals in Scotland like the Royal Hospital for Sick Children in Edinburgh, which still lies empty after eight years.

The other devolved administrations and the London mayoralty have followed suit and the pressure may well lead to the whole country downing tools. If this happens the effects will be far-reaching. As a high proportion of the building workforce is Eastern European there is no guarantee that they will be picked up again. Many have already gone home to beat the lockdown and those that stay may not have much loyalty to Britain once they have been laid off for months. Even those with settled status may decide to leave. We face a huge skills shortage without them.

It is potentially disastrous for me if it delays the milking parlour and new cattle sheds. The work is all outside and social distancing can be maintained. Every week lost in the summer is critical; many jobs like laying concrete won't be possible once we lose the good weather in the autumn. We need to start milking in February.

But it is far worse for the many small builders around here facing an uncertain future. In the countryside farming and building are practically indivisible. They cannot understand why farmers are allowed to drive tractors while they are not allowed to drive diggers. The argument that construction workers are sitting next to NHS workers on public transport in London or Glasgow seems absurd here in Galloway. It doesn't take much for self-employed builders to go bust at the best of times. Many of their workers are also self-employed sub-contractors who cannot just be put on furlough. The mixed messages coming from different capitals has led much of the building supply chain to close down now, which will now bring construction to a grinding halt anyhow for lack of materials – one local firm closed then re-opened and may be closing again. The Chancellor is now under extra pressure to bail out the self-employed.

This ruckus over building sites exposes the fault line running through British politics. Capitalists like Boris Johnson see the critical importance of the building industry to the wider economy. Socialists like Nicola Sturgeon and Sadiq Khan assume that the Big State has taken over and will always have a magic money tree to shake. But it is vital that the former view prevails to ensure a swift recovery. It takes robust leadership in times of national emergency for politicians to tell people to go to work when their families are telling them to stay at home, especially when they have already stopped once. But they must all pull together in all four nations to make sure that any lay-off is as short as possible. We can't build our way out of recession if we have no builders.

* * *

Calving is one thing that has not been halted by Covid-19. Thankfully none of us has succumbed yet as the chances of finding someone else to come and work the calving jack are remote in the extreme. The contingency plan is 'keep calm and carry on'. We can always use the old remedy and drink colostrum, which is one thing that isn't in short supply around here. As I stand in the calving pen watching a cow eating pink and purple spaghetti I ponder when it was that human mothers ceased to eat afterbirth. Presumably when nutrients became more readily available and we no longer had to hide births from predators. From too yummy to munch to too posh to push in how many millennia?

* * *

When I was the intelligence officer of my battalion as a young captain in West Belfast in the late 1980s I always had a copy of Sinn Féin's newspaper *An Phoblacht* on my desk. I am reminded of it every time I come up against Scottish nationalists on Twitter, or read their newspaper, *The National*. The similarities between the two narratives are chilling. There is the same post-modernist approach to the truth: fake news, fake history, fake statistics. The SNP Twitter trolls inhabit their own little world where Scotland is a colony of Britain. Their own ugly brand of nationalism is portrayed as 'civic and joyous' and they excuse it by calling unionists 'Brit Nats' and claiming, in much the same way as Gerry Adams did, that we are forcing our own brand of nationalism on them.

Britain and British are only ever used in a pejorative sense. This is deeply shocking to me after a twenty-year career serving in that most British of institutions, the army, where the synergy of the United Kingdom is most evident. The four home nations generate armed forces that are worth so much more than the sum of their parts on the world stage. I remember my first overseas tour in the Falkland Islands where the local population were so proud to be British; then Hong Kong where the people we colonised are now desperate to become British citizens. I helped to train the newly unified Mozambican army in Africa, formed from the two opposing sides in a bitter civil war, the former child soldiers of Renamo and Frelimo. Mozambique subsequently joined the British Commonwealth despite having never been a British colony. In Sierra Leone, once our richest colony, life expectancy had dropped to thirty-seven as a result of the civil war. In a bar in Freetown I met Kate Adie, the BBC's chief news correspondent. She said she was amazed at how many Sierra Leoneans had expressed the hope to her that Britain was going to re-colonise them.[36] But my sharpest

---

36. This didn't make it into her broadcasts however!

memory of the way in which others see Britain was in a sandstorm in the Iraqi desert. We had crossed the minefield into Iraq the night before and all through the night demoralised Iraqi soldiers had appeared out of the darkness with their hands up. As day broke it was clear that we had been swamped by prisoners. I stopped the advance and ordered up a field kitchen to give them a hot meal. They had been under intense bombardment for nine days during which supply of rations or mail had been impossible. They all said that their officers had told them, 'Surrender to the British if you can. You will be well treated.'

All these foreigners must all be very confused by the notion that there are those within Britain who see her as some sort of evil empire and wish to break her up. It is very evident that many of the more extreme Scottish nationalists have never set foot outside the UK and have no idea how Britain is seen in the world. And a generation of education, in which our history is either missed out altogether or shown in a bad light, has allowed false narratives to develop.[37]

* * *

Covid is exposing the fault lines in our constitution. Imagine this really was a war and we had the Prime Minister saying 'We must fight them on the beaches' immediately followed by the First Minister of Scotland saying 'The beaches are out of bounds. Stay at home.' The pandemic is devolution's first big test. It has had some benefits as well as negatives in 'peacetime' but it now risks damaging our ability to cope with a national emergency, both in Scotland and in the rest of the UK. It had been agreed that we would fight this 'as one United Kingdom' but it is now evident that the national effort is being hampered by Nicola Sturgeon not sticking to the script agreed in COBRA (of which she is a member) and it is adding unnecessarily to the fog of war.

Sturgeon is under intense pressure in the aftermath of the Salmond trial with a looming public inquiry into what she knew

---

37. One example is the story that Churchill sent tanks into George Square in Glasgow to put down the 'Red Clydeside workers' protest in 1919. It never happened.

and when about Salmond's 'inappropriate behaviour' with women. This partly explains her determination to keep a narrative of decisive leadership on the front page. Sometimes she has rushed out of COBRA to have her say first. And her insistence on differentiating everything has meant that her press conferences have sometimes contradicted the Prime Minister's to the chagrin of many in Scotland as well as Downing Street.

BBC Scotland has not helped, talking about 'the United Kingdom and Scotland' as if independence has already happened. Nor has mutual antipathy between Johnson and Sturgeon. She has made him the bogey man in Scottish politics since the Brexit referendum. He has perhaps not helped by calling her 'That Wee Jimmy Krankie Woman' behind her back.

Beneath the posturing the First Minister is actually not having a good war. The key worker strategy was badly botched. Sturgeon ducked responsibility and delegated it to local councils. Consequently there has been confusion and many critical functions not working in some areas. The list is much clearer and more comprehensive in England. This is something that would have been better handled at a UK level.

We are now in a render-unto-Caesar situation. People in Scotland are starting to ask, 'Are we taking orders from the organ grinder in London or the monkey in Edinburgh?' In Scotland most businesses are now closed while in England the message is 'keep calm and carry on' if you can, with precautions. The former undermines management in the latter who are trying to motivate their teams to keep working. These mixed messages drive wedges between employers and employees. There is more than a hint of socialist dog whistling in the narrative of wicked capitalists putting their workers at risk.

More seriously, critical financial aid in Sunak's rescue package is not reaching where it is needed because Holyrood has insisted on differentiation. This is a toxic issue for the nationalists. It is dawning on Scots that an independent Scotland could not afford a bailout like the Chancellor's. The tourism industry is asking why self-caterers are excluded from receiving aid unlike English and Welsh counterparts. One consequence is that farming businesses like ours, which rely on self-catering tourism at this time of year to pay for seed and fertiliser, have lost the liquidity to plant this year's crops.

CHAPTER 9

Lockdown

There is very little good to report at the moment but we can celebrate the news that Faith, Hope, Charity, Grace, Humility, Patience, Prudence and Chastity have had their lives saved as a result of coronavirus. These are our new hens (Chastity drew the short straw at the naming). They were due to meet a sticky end, being coated in barbecue sauce in ready meals, after their egg laying curves had started dipping at the chicken farm up the road, when the Quartermaster-General came back from an unfulfilled visit to the supermarket and decreed that we need to get back into egg production pronto, 'And while you are at it, you need to dig over the vegetable garden so we can get those seeds in.' Now that we have closed the B&B and the holiday cottages until further notice it is goodbye Basil and Sybil and hello Tom and Barbara for us.

We gave up keeping hens when the children left home but now that they are back with us – when they are not conducting video-conference calls from the drawing room – the eggs are very necessary. I had forgotten the calming influence of chickens. If I am ever in a coma they should play me recordings from the hen run of their contented mewing and clucking. I think I must find it comforting because my earliest memories are of going to collect the eggs at my grandparents' farm with Mary, a wonderful eccentric from County Mayo, who had come to live in a caravan as a Land Girl in the war and never left.

Memories of our children's early life are also closely woven with chickens. They started in an army quarter in Wilton. The chooks arrived and promptly flew off as I had forgotten the old keeper's trick of clipping one wing. So retrieving them from other people's gardens was a very good way of meeting the neighbours. Later we had Pekin bantams, Napoleon and Josephine.

Napoleon had a close shave when my late sister Annabel came to lunch with her border terrier. The cockerel had a hell of a scrap protecting his family from the dog and ended up plucked, half oven-ready and badly chewed. After forty-eight hours in intensive care in the warmer drawer of the Aga he pulled through only to die the following year when he attacked his reflection in the water trough and drowned.

I mused about getting a rooster to keep the new hens company. I love hearing the sound of a cockerel calling above the dawn chorus, competing with the cock pheasants cock-cocking and beating their chests; it's like adding a classical saxophone to the orchestra. And I like the way they strut about asserting masculine values. The Quartermaster-General was dismissive of this idea and said that the hens didn't want to be bothered. I was rather hurt.

One member of the family who needs to be isolated from the chickens is our Norfolk cross Lucas terrier, Pippin. Chickens must be the stupidest members of the avian world. Any intelligence has been domesticated out of them. Wild birds are much brighter and even display a sense of humour. The crows love teasing Pippin by flapping off to land just out of her reach and repeating it several times causing her to flare across the beach barking.

* * *

At last our bank account is getting some infusions of cash. 'Rishi's millions' are providing some compensation for the loss of holiday cottage income. Although for every ten pounds, the SNP administration in Edinburgh seems to be trousering three for their own pet schemes and so our competitors in Cumbria just across the water have more money to stay afloat. The last jaunt before lockdown was to Yorkshire to speak at the Bedale Hunt end of season dinner. I was reassured by how the farmers spoke very highly of their local MP, one Rishi Sunak. After being elected he was put on a crash course in farming by some of the elders of the Dales. They say he passed with flying colours. But the highest praise came from my old friend Trots, recently retired District Commissioner of the Bedale Pony Club. The Chancellor of the Exchequer is apparently a top Pony Club dad who 'mucks in and is always there at the end of a rally to put the jumps away'. You don't get a better reference than that.

***

Pandemic pain is now being passed right back up the supply chain to farmers. Social media are full of heartbreaking stories of milk being poured into slurry tanks on British farms because processors that normally supply the catering trade in bags are no longer picking it up as they can't bottle it. This is every dairy farmer's worst nightmare. It makes us anxious about our decision to get back into dairy farming but I am holding fast to the first principle of war, remembered from my first day at Sandhurst: selection and maintenance of the aim.

The beef price has also plummeted as the supply chain can't cope. The abattoirs and processors have slowed their throughput. But there is always someone worse off. My friend Simon Duffin emails to see if I can write an article about the plight of trout farmers. Supermarkets have dropped selling trout portions to streamline their operations and fishing lakes are shut (perhaps unnecessarily?) so aren't restocking. Millions of trout may have to be killed and tipped into landfill. Meanwhile empty supermarket shelves show that there is no let-up in demand for food. In normal times 20 per cent of food is consumed outside the home and the switch to home consumption has provided a challenge for the supermarkets – but also a golden opportunity. Special offers have been withdrawn and prices have been nudged up. Mince that was selling for £2.99 per kilo in Tesco is now selling for £4.99.

You would think that at least farmers might receive more for their produce, given the constrictions on supply, but in fact the reverse has happened. Beef farmers have had to swallow another ten pence per kilo drop in the price this week. Milk is around seventy pence per litre in the shops while the spot price being offered to farmers has fallen to as low as fifteen pence in some parts of the country – for a product that is instantly digestible when it comes out of the cow. The farmer share of the meat retail price was already at an all-time low before the pandemic; the gap has widened since.

These effects on the supply chain could have been predicted but the government has adopted the laissez-faire approach of leaving it to the supermarket cartel to ensure that we have enough food. Competition rules were suspended to allow them to work together. They appear to have responded by profiteering on an unprecedented scale. Two of them immediately arranged for the import of cheap beef from Poland, which has sent British farm gate prices even lower. It is alleged that they waived their normal quality assurance standards to do this. Polish beef is banned in a number of EU countries after it was found last year that an abattoir had been processing sick animals.

Ministers have gone out of their way to praise supermarket workers for carrying on working (for which many have received bonuses) as if food is grown in their warehouses. Farmers, many of whom have no choice but to carry on working while ill, are angry that food companies and supermarkets are profiteering at their expense. DEFRA has been too slow to re-allocate labour or even request military assistance to drive milk tankers and backfill food production workers. Last week DEFRA Secretary George Eustice was shamed into putting out a statement thanking British farmers, but the fact remains that livestock farmers are now losing money through his department's failure to plan for failures in the supply chain preventing the flow of British food from field to fork.

April 2020

Since moving home a decade ago I have learnt to mark the progress of the seasons by the birds, and particularly the long, aching search for signs of spring. It is striking how similar each year is. The human world may have been plunged into lockdown by a global pandemic but the avian one is doing its thing as usual, if anything with rather more freedom. As ever, it started on Valentine's Day with a pair of shelduck appearing on the foreshore below the house. Then, as each morning new melodies were added to the dawn chorus, I noticed other old friends, or probably the great-grandchildren of old friends re-appearing in familiar haunts like the wagtail yo-yoing down the road in front of me on successive mornings until he disappeared. I hope he wasn't had by the sparrowhawk. Or his pied cousins appearing bang on cue as we started ploughing to flit back and forth over the freshly turned furrow looking for seasonal delicacies like housewives in a re-opened delicatessen. Then as the cherry trees behind the house started showing off I saw a pair of bullfinches darting through the branches. Finally, earlier this week, the sweetest moment of the year came as I was moving the electric fence to let the new dairy heifers on to their next patch of fresh grass. Out of the corner of my eye I glimpsed a familiar flash of midnight blue and white and there they were, the first swallows of the year – appropriately enough on Easter Sunday – confirmation that the vernal happenings to date have not been a mirage and all is well with the world. Swallows and cattle have a symbiotic relationship. The densest concentration of insects on the farm is always around the cattle and the swallows are there now most days hawking backwards and forwards over the cows rebuilding their strength for breeding. And the muddy patches are critical for the provision of nesting materials. Numbers have been down in recent years when we have done the count so hopefully last year's good breeding season will mean that a few more reappear this year from their winter in the glorious veldt of the Eastern Transvaal.

And yes, you did hear me say dairy heifers. I know that it is really happening as they have arrived. Unloading them off the back of a lorry from Southern Ireland is a big moment. One small problem is that they have arrived well in advance of any staff to look after them so Muggins here is looking after them. The whole place is being laid out with cow tracks and paddocks to enable them to graze in mobs and move onto fresh grass regularly. It is a step change from the old way of doing things by 'set stocking' the fields with the right number of cattle to stay there all summer. It is taking a while for my eye to switch from appreciating the chunkiness of beef animals to the thoroughbred lines of dairy heifers. I have a pang of regret that soon we will no longer be able to enjoy summer days watching suckler cows cudding lazily while their calves frolic around them but I am looking forward to seeing dairy cows here again after forty-four years. There is a logical symmetry to what we are doing: shortly after we entered the EEC my father got out of dairying and now that we have exited the EU I am going back in again.

We are also cutting back on the arable side of the business to create more grass and this has some pleasing consequences. The 'milking platform' of 600 acres of grass sounds industrial but is in fact restoring a parkland landscape and now that the eye is no longer distracted by the ugly scars left by arable cultivation one can better appreciate the copses, walls and hedges. One morning last week there was a rainbow with its end resting on where the new parlour will be. I hope it's a good omen.

* * *

I am reviewing *Feeding Britain: Our Food Problems and How to Fix Them* by Tim Lang for the *Telegraph*.

There hasn't been a better month in which to bring out a book about food security since the middle of the Second World War. The food rationing by the supermarkets is a classic example of what Tim Lang, who is Professor of Food Policy at City University, calls the 'Leave it to Tesco Reflex' by government.

The numbers are stark. We are currently reliant on 10,000 food containers (50,000 tonnes) per day coming from the EU. British farmers are being appreciated during the pandemic but, much as it might grate to do so, we should also thank Ursula von der Leyen for insisting that European countries keep borders open during the pandemic for the free movement of goods. It is literally saving our bacon. All things considered, it is amazing that we have not started to go hungry. Lang records estimates that, in the hauliers' strike of 2000 that caused the Blair government to wobble, we were two to three days from disaster. It is likely that experience influenced no-deal planning and we have the Operation Yellowhammer[38] planners to thank for keeping shelves at least part full; Lang describes just-in-time supply as being where 'the warehouse is actually the motorway'.

---

38. The contingency plan for a 'no-deal Brexit'.

The book offers a very comprehensive overview of what Lang describes as our 'food system', which encompasses diet, the environment and things like our over-reliance on plastics as well as the supply chain. He has a pinkish view of life. It is hard to square his emphasis on the use of food banks – a relatively recent innovation – with the fact that food has never been so cheap in real terms. Since Queen Elizabeth II has been on the throne average household expenditure on food has fallen from around 40 per cent to under 10 per cent, something that has never before happened in one reign. But it is hard to disagree with his thesis that our imperial past has made us dangerously complacent about food supply and over reliant on imports without asking too many questions about the food or the effects its production is having on the environment elsewhere. And no one could describe our population as thin. Lang details the 'massive diet-related, NHS bankrupting food culture' and he explicitly links the chronic burden caused by obesity and diabetes with the fact that we eat more 'ultra-processed' food than anywhere else in Europe. He is one of several authors to assert lately that we need to eat more simply and cut out the sugars and starches. One startling quote is the NHS Chief Executive's acknowledgement that the costs of treating the results of excess eating were more than the Home Office allocation for the police in 2016. I suspect that they have increased since and many of those dying from coronavirus with 'underlying issues' will fall into that category.

As a former Lancashire hill farmer, Lang understands agriculture. He calls for a Great Food Transformation and writes that 'we should start now or be forced to do it in a crisis later'. Little did he know that the crisis was just around the corner. His recommended approach is to work out our food requirements and then design the farming system we need. He believes we should aim for 80 per cent self-sufficiency. He takes a Remainer view of Brexit but undoubtedly we will now have more freedom to do that. His plan to make beef and dairy grass-fed and switch the cereal ground this would free up to remedying our deficits in fruit and vegetables will meet with resistance from some farmers but must be the right way to go. And his insistence that farmers and fishermen must be given a fairer share of the Gross Value Added to food than their current 8 per cent in order to make those industries profitable and attractive again will raise a cheer in the countryside.

Lang and others have been 'prophets in their own country' for three generations. I hope they will now have their place in the sun. Few can doubt that the Agriculture Bill presently going through parliament will have to be completely re-written from first principles post Covid-19 with food security and a national nutritional policy at its heart.[39]

* * *

If 'a peck of dust in March is worth a king's ransom' then what is a whole April of dust worth? Weather lore doesn't tell us because it is unprecedented, here in Galloway anyway. And, of all the Aprils for it to happen, this one has been heaven-sent as the whole place has been covered in diggers and dumper trucks laying seven kilometres of cow tracks. Most years they would have been axle deep in mud. The tracks seem more like embryonic motorways. The quarry looks more like the Big Hole at Kimberley every day when I go to enquire anxiously of Stevie the digger driver whether he is still hitting rock. We found a rich seam of clay with shale running through it and once carefully graded, cambered and rolled into a pleasing rounded shape then baked in the sun, it has formed what looks like a durable surface. We will see. They say there are three things to remember about road construction: keep the water off, keep the water off and keep the water off. As I write, the first rain for a month is pounding on the roof of my cabin so I will be off to search for incipient potholes when it clears. At least it is laying the dust and there is briefly that blissful African smell of rain on parched land that must trigger some primal contentment within us.

For now the immediate concern is bulling the heifers. They are munching their way round in one big mob like a herd of buffalo across the Serengeti. Judging by the amount of sapphic activity they should stand for the bulls when they go out on May Day. One of the unanticipated pleasures is moving the electric fence each morning and hearing them pulling at fresh grass while the birds

---

39. It has not been so far.

harvest insects on the freshly manured pasture behind them. The estate has come full circle and I am frequently stopped by older members of the parish – at a social distance – and told how nice it will be to see dairy cows back again.

* * *

Another middle-aged rite of passage. Eric Anderson, my old headmaster has died.

I once attended a parents' seminar at my children's school where we discussed the challenges of educating teenagers. The subject turned to drugs and the master leading our syndicate gravely stated the policy, which was zero tolerance: one puff and you are out. 'That is very interesting,' said I. 'I remember a boy a couple of years below me in my house who was caught smoking a joint. The headmaster took the view that it was a first offence and that the boy was fundamentally good and didn't expel him. And now he's the Prime Minister – the boy not the headmaster.' Probably seven out of ten headmasters would have taken the path of least risk and expelled the boy (one Cameron minor). Of the other three, two would have been wet as water and heading for expulsion themselves for lack of grip. The last was Eric. We probably have him more than anyone else to thank – or blame, depending on your point of view – for the Cameron premiership. Certainly without Eric's humanity and confidence in his own sound judgement, Cameron would not have progressed to Oxford and the leadership of the Tory Party at the age of thirty-nine; we might not have had the recovery from Gordon Brown's stewardship of the public finances and in all probability, though things didn't turn out quite as Dave intended, we would still be in the EU.

The obituaries predictably focused on the famous people Eric taught. When those tedious Guardianista journalists publish articles showing networks of influence implying masonic handshakes and corrupt establishment conspiracy theories, one of the great and good with a better claim than most to be at the centre of the web was Sir William Eric Kinloch Anderson KT, one of only sixteen Knights of the Thistle, Scotland's highest honour. I wish they could have met him; a humbler man with less 'side' to him would be hard to find. The shy kilt-maker's son with the mild rectitude of a Moderator of the Church of Scotland was the embodiment of the idea that the most essential prerequisite for greatness is humility.

I had no idea he was ill. I last saw him at the launch of my book *The Enigma of Kidson*. He stood grinning in front of the lectern, a little greyer and more stooped than I remembered but a very fit looking octogenarian listening to David Cameron, who had kindly agreed to make a rare public outing to say a few words ('My name is David Cameron. I used to be Prime Minister, now I'm the warm-up man ...'). Cameron used the opportunity to thank Eric publicly for not expelling him when he richly deserved it.

Headmasters of Eton are dissected at dining tables in London clubs and country houses with the same interest that managers of football teams are analysed in sergeants' messes and working men's clubs. I can remember six of them being discussed and Eric is the only one to have earned universal approval. My first memory of his arrival was that the snap judgement of the boys that he was 'cool', and the fact that he was referred to by all as Eric, a subconscious acknowledgement that he was human like the rest of us. That certainly was not the case with the man he replaced, the terrifying McCrum, who was to me what Dr Arnold was to Flashman and had already beaten me twice – eight of the best for a fifteen-year-old, something that would probably earn him a stretch in Wandsworth now. This involved a humiliating ritual on the beating block in Upper School, and is recounted in all its gory detail in *Kidson*. To my great relief, Eric's first act was to consign the block to the museum.

I would like to be able to say that I was in his essay group along with Boris and the other swots but I can't say that our paths crossed much at school. Alas, when they did, it was invariably 'on the Bill' when I was before him for some misdemeanour. I remember coming away each time thinking that he had dealt with a delicate situation, which he clearly found as trying as I did, with commendable good humour and an attempt to make it as short and painless for us both as he could. It was an early lesson in lightness of touch and the disarming power of not exerting power and using charm and persuasion instead.

After I left I saw him briefly at the quinquennial gatherings of the class of '82 (front runner Johnson major) where Eric would give one of his beguiling, self-deprecating speeches that effortlessly extracted donations from wealthier classmates for a scholarship fund that ballooned under his stewardship and stands as a powerful tool for the advancement of social mobility for the nation's gifted boys, much to the chagrin of the leftists.

I only came to appreciate Eric's greatness fully when I came to write the book, a biography of my old tutor, Michael Kidson. Eric had given the address at Kidson's funeral, which I had been unable to attend, and a friend had sent me a copy. It moved me deeply and encapsulated Michael in two sides of A4 better than I could hope to in a whole book.

His words from the pulpit summed up Michael Kidson but also nailed what it is to be a great teacher:

> He may have been appointed as a teacher of History, but actually he was a teacher of boys. And that is much more important ... for him Eton was not a place, not buildings and rackets courts and fields and colleagues and community. Eton was you: 'his' boys. His relationships with you were what Eton meant to him. Your loyalty and continuing friendship were his reward, and he died – I am certain – happy in the knowledge that he had mattered and had made a difference to so many of you.

They could have been written about Eric.

May 2020

It feels wrong to write this when there is so much misery, but I think the lockdown will be etched on my memory as a glorious spring when we have been able to appreciate the wild flowers and the birdsong without being troubled by the outside world at all. The call centres are in lockdown, the civil servants have been furloughed and travelling salesmen no longer happen to be passing our door. We would normally have been rushed off our feet by now dealing with questions, requests and complaints. But, much as we enjoy having people staying in our holiday cottages, we have been guiltily relieved to be getting a proportion of the money for nothing. No doubt there will be a nasty sting in the tail when the government needs to replenish the coffers.

The place is looking so bonny with sunlit crab apples competing now with bird cherries and hawthorns that it seems wrong to pray for rain. But we are beginning to fear for grass in the new dairy paddocks that we planted into dryish seed beds and rolled in tight to preserve the moisture. Establishing it was a challenge. The ground was so hard that we had to plough some of the heavier land just to get some tilth rather than direct drill into the dead sward. It has germinated but we are rather wishing it hadn't until it rains. I thought aloud about irrigating but was quoted £100 per acre per inch of water and thought better of it. On a brighter note, I heard the first cuckoo yesterday, which bolstered my morale. I didn't hear it at all here last year. And the swallows are back in profusion. There was only one pair for the first week and we waited like anxious parents in an airport arrivals lounge to see if more had survived the trip. Now the yard is regularly dive bombed by squadrons of them.

The lockdown has also reduced the number of walkers, and this too is something of a relief. The regulars from the village have kept doing their circuits. And it has been amusing to watch the social distancing interpretations evolving as neighbours walk together. The new cow tracks for the dairy operation have proved irresistible to them, as I feared they would, but it might be a different story when 650 cows have walked down them leaving them coated in 'skitters' as we call that other word usually suffixed with an 'e' in the Scots dialect.

The new dairy heifers now have bulls running with them so that they calve in February – a dozen of them all in together so that it is like watching a herd of wild buffalo complete with sound effects. There has been the odd scrap but they seem to have established a pecking order quite quickly. The younger bulls appear to be doing all the work chasing heifers round the paddock while the old bulls sit quietly in a corner conserving their energy – but then one of them rises, ambles over to a receptive heifer, sees off the young teasers and does the business. That's one more ready for the parlour in February hopefully.

Meanwhile in the silage fields the larks are busy nesting. I am praying that they are fledged before the mowers arrive in the third week of May. Our new regime should suit them well. The cattle graze each paddock down low then don't return for about three weeks. With an eleven-day incubation period this gives the larks a good chance. In fact I am sure they have evolved by nesting on freshly grazed pasture for that reason.

***

The drought continues and grass-anxiety is on every farmer's lips. It is interesting to see how simulating nature with our new grazing regime is helping us cope. Where we are mob grazing small paddocks with 250 animals the grass grows back strongly where they have been. Where we are set-stocking fields with the last of the beef cows and calves it is brown and shot.

The highlight of this spring's birdsong has been more cuckoos calling than I have heard for years. Whether this is due to our success with the Larsen traps or self-control by the Maltese we can't tell, probably a bit of both. But it is when birds squawk strange oaths at each other that spring is at its most interesting. Yesterday I was startled by the avian equivalent of two fishwives hurling insults across a street when a peregrine lumbered over my head. I know, peregrines don't usually lumber, but this one had a bird almost as big as herself in her talons and she was yawing like a helicopter with an overweight underslung load as a crow attacked her, attempting to steal her prey. The two disappeared into dead ground and I ran to see the fight. It looked as if the falcon had gone to ground to mantle over the kill but next I saw the two of them spiral up in a dogfight. The air was thick with black feathers and the peregrine was clearly besting the bigger bird, which flew off in disgust. By good fortune I had my binoculars and watched as she circled above the fields. Peregrines are never allowed through other birds' airspace unmobbed, especially at nesting time, and sure enough she set off a chain reaction as a buzzard flapped out of the old beech trees by the smithy and wheeled up to see her off, mewing confidently. But the peregrine, miffed at losing her prey, took out her anger on the buzzard and stooped on it several times. Had she made contact with such a big bird it could have been suicidal but she deliberately missed by a faction of an inch so that the buzzard had to fold its wings and crumple like a shot pheasant each time before recovering as the falcon hurtled past. I watched fascinated as the two then separated and flew high into the ether.

Going over to look for the dropped prey I expected to find a dead pigeon or jackdaw but I found a very dazed feral dove from a nearby steading, which flew off unsteadily when it saw me. I rang Greg, our go-to expert for all matters ornithological, to tell him. I asked him why the pigeon had not been killed. He explained that in late May the parents catch birds carefully in their talons and carry them back to their young to teach them to hunt. That was one lucky pigeon.

One silver lining of the global pandemic is that our children have been stranded here, working from home and therefore able to witness the spring flowers for the first time since they were packed off to school. We have enjoyed watching them marvelling at the rhododendrons and azaleas in all their vulgar glory. Best of all have been our bluebell woods. They get better every year and the carpet of blue fringed with dazzling white drifts of garlic has been spreading gradually into new parts of our woodland as we have thinned bits and raised the canopy to let the light in. It amuses me that, for the green establishment that now hogs the airwaves, bluebell woods are rare examples of our wild environment. Nothing could be further than the truth here. Where I have re-wilded bits they have been colonised by a scrub of willow and birch. We had far fewer bluebells here when I was a child, but then as we harvested the conifer woods, planted by my grandfather under the influence of a timber hungry post-war government, and my father re-planted with native hardwoods, the conditions were gradually created for the hypnotising fog of violet blue that we have now. It is a policy I have continued. The joke is that the bluebells thrive on a symbiosis with oaks and beeches in particular. The latter, though they were here before the last ice age, are not considered indigenous and thought to be very non-PC by the apparatchiks of the Green State who oversee planting schemes. What a good thing it is that they have been cheerfully ignored.

* * * *

June 2020

The first cut silage is in the pit, something in the larder to see us through the winter. There wasn't much of it but it was bone dry so in the annual speculation about how much bigger or smaller it would look if the moisture level was different we consoled ourselves, without any scientific evidence, that the dry matter 'probably' equated to what we get in normal years. We'll see. It was so dry that, even with the contractor's new five-tonne roller following the buck rake, the silage was soft and bouncy when we walked over it to put the sheets on. I have never seen it so dry.

Silage chat on Twitter led to a sharp riposte from Robin Page. He tweeted 'I'm sorry Jamie but I don't think you should be making silage in May.' It brought me up short and led to a bout of soul searching. Robin is someone I like and respect and Lark Rise Farm, his oasis of wildlife in the arable desert of Cambridgeshire, is a model of conservation. It's also true that whereas the chemical-soaked monocultures have destroyed much of the wildlife of that part of East Anglia, the multi-cut silage regimes of grass farms here in the west threaten to do the same – particularly those feeding zero-grazing operations or anaerobic digesters. It's a finely balanced argument here where we try to farm in a wildlife sensitive way while trying to make a profit, and the best is often the enemy of the good. Until this year we have usually silaged around Ascot Week towards the end of June and then again in August. That has given the ground-nesting birds and hares time to breed while ensuring enough silage of a reasonable quality to feed the beef cattle. But converting to dairy brings wildlife opportunities and threats. Dairy cows need better quality silage to augment grazed grass when they start producing milk in February, and that means cutting a few fields of leafy spring grass before the heads have set, which means before the end of May, and then again later in the summer for the dry cow silage, which doesn't need to be as good. This will inevitably be a threat, although I was relieved to see good numbers of first brood larks on the new mown fields last week.

This year there was a new man on the mower. I asked him to mow from the middle of the field out to the hedges so that wildlife had a good chance of escape. He assured me that he had done a season in New Zealand where such practices are mandatory.

Why not here? Depressingly the extra badger setts we are seeing every year mean that, whatever we do, an ever greater proportion of birds and leverets will become badger food. I don't think the Kiwis would tolerate that either.

* * *

It's a typical night in cyberspace above Edinburgh; a Twitter pile-on is raging against a retired academic named 'Historywoman'[40] for daring to speak the truth about Nicola Sturgeon, her cult of fanatical separatists and the gradual erosion of free speech and liberal values in Scotland. If we survive the Sturgeon terror there is a film to be made about the brave women who have kept the flame of democracy alive, a sort of *Calendar Girls* meets *The Gulag Archipelago*. Many of them appear to be grandmothers horrified by the perverted future being created for their grandchildren. I have an image in my head of these doughty defenders of the Scottish Enlightenment – Maureen Johnson, Rosie Lickspittle, Aviemore, Georgina, Mrs F and many others – as Miss Marples coming back from bridge, completing *The Times* crossword, pouring themselves stiff drams and sitting down at their keyboards to put in a shift defending the union into the small hours against the trolls.

I first became aware of them in an early foray on Twitter when I was shocked to the core by a post from 'Queen of the Flumps' (the Flumps appear to be her dogs, on whom she clearly dotes). She had posted a video of herself addressing her nationalist MP through her car window, evidently captured by her smartphone on the passenger seat. She asks him why he has yet to hold a surgery for his constituents. He attempts to intimidate her by warning her about 'the stuff you have been putting out on Twitter' and telling her she is 'on our database.'[41] Intrigued, I researched her Twitter

---

40. Professor Jill Stephenson, Professor Emeritus of Modern History at Edinburgh University, later to become a friend and mentor.

41. You can still see the video online. The MP, Alyn Smith has never been investigated. Aside from the creepiness, it is almost certainly illegal for a political party to hold such a database for the purposes of identifying political opponents. And if it isn't, it should be.

output, interviewed her and found that she is a highly educated professional whose only crime is to be critical of the SNP and to ask her MP when he is going to have a surgery. (His idea of a surgery appears to be to swagger round the streets with the heavies claiming that he is conducting an 'open-air surgery'. You can see this on his Twitter timeline.) He denied telling her she was on a SNP database when asked by her followers if he'd said it. She then asked him to retract calling her a liar. 'I told him that he knew he'd said it, I knew he'd said it and I had proof that he'd said it. I gave him one hour, he continued to deny it so I uploaded the video clip.' She tweeted the video recording of him saying all these things with the words, 'OK Twitter, you decide.' And BOOM!

Except that it wasn't boom. If a Tory MP had intimidated one of his constituents in such menacing terms it would have been front page news and the Speaker would have become involved. There was an outcry on Twitter but the real story is how Scotland's journalists have behaved. They have ignored the story. Worse, two senior journalists with nationalist leanings actually supported the MP by trolling her, and engineered a troll pile-on by SNP supporters by accusing her on Twitter of being a 'known stalker' and 'having mental health problems'. The cybernats duly piled on. 'They not only threatened to rape me, but threatened to kill me and kill my dogs in front of me.'

She still sounds very upset. But mostly she is really angry, as I am, that Scotland is rapidly becoming like 1930s Germany. I attempt to pitch the story to newspapers and radio stations but none of them seem to want to highlight SNP wrongdoing. It is really shocking I grew up thinking of Scotland as a land of milk and honey where there was genuine kindness wherever one went. The adjective most used by the feisty women on Twitter when calling out SNP evils is 'rancid'.

***

'Oh man! US farmers would never put up with all this c**p!' I can picture her now, the young American cowgirl doing an internship here last year. She was shocked that we have to tag all our animals at birth and then record every antibiotic administered throughout their lives so as to ensure the complete traceability and food safety of British beef. I remember the look of incredulity on her face as I pointed out the grey hairs on my head caused by mislaid cow passports. And that is just one example of why the US trade deal isn't as simple as some are making out.

The government has just voted down an amendment to the Agriculture Bill saying that it was unnecessary because they had already committed to banning hormone-treated beef from entering the country. That salved the consciences of the majority of MPs. But British farmers know that they would go on burning the candle at both ends to keep very onerous records while beef produced without the same traceability floods in. Whatever the assurances and the red lines there simply aren't the record-keeping systems on US farms or the controls in the supply chain to ensure that imported food is produced to the same standard.

Despite having reluctantly voted Remain, I have some sympathy with the frustration of politicians who want to do free trade deals around the world in order to reduce prices for British consumers and create opportunities for exports. I think the farming unions have been using the wrong arguments. Americans have not been dying like flies from eating chlorinated chicken. Our unions have ignored the Prime Minister's warning in his Greenwich speech that 'we will not be fooled by mumbo jumbo' and doubled down on weak arguments.

BUT there are some pretty big circles to square if we are to do this trade deal without destroying a large chunk of our agricultural base, and with it our landscape, environment and rural communities. The government needs to be far more diligent in war-gaming the scenario of US meat, grown faster and cheaper with hard to detect hormones and GM feed – undercutting home-grown products and destroying British family farms. It would seriously undermine the union as the effects would be felt disproportionately in Scotland and Wales. The SNP have been adept at developing a narrative of Tory betrayal and this will be a big issue in key rural marginals in next year's Holyrood elections.

The rot has already started. It is not just us getting out of beef. One friend is selling his cows and planting his hill farm with trees. He is laying off three men and will no longer be spending £600,000 per year on agricultural supplies in the local market towns. That's also less moorland habitat for rare curlews and hen harriers.

If we are to do a trade deal with the Americans – and we probably should, not least to put us in a stronger position with the EU – we need a clear vision of an end-state that will sustain a viable British beef industry. We should start with labelling. Currently meat can be labelled as British if it is packaged here. That needs to change. If farmers have to keep records that ensure full traceability, the birthplace of the animal needs to be on the supermarket label. We also need much clearer differentiation between British mainly grass-fed beef and US feedlot soya-and-cereal-fed beef. They are completely different products. The former is much better for health and the environment than the latter. One's public relations is undermined by the other. Our farming industry has been reluctant to differentiate between grass-fed and cereal-fed to avoid upsetting the big beef finishers, but they must now do so or they risk being blown away.

And we must create the conditions that will allow British farmers to compete on quality. That means more small abattoirs. And we must do more to boost exports. The Chinese would love to eat more British beef. They have discovered that beef protein makes their children grow taller and during the pandemic they have learnt the importance of meat and dairy for the immune system. They want to build abattoirs in the UK. They should be allowed to. Whatever our quarrels with the Chinese at the moment, if we don't sell them our beef there is only one loser. British farming is on a knife edge at the moment and there is no sign of the kind of dialogue that is needed between government and farming unions to enable us to find the right solution.

July 2020

The drought is a distant memory and the monsoon has arrived early. It normally waits until we are about to start harvest. This has thrown our grassland management plans. Back in the spring we stood listening to Brian the contractor shaking his head dolefully and announcing that the ground was so rock hard that he couldn't direct-drill and needed to do a shallow plough just 'to make some soil'. (Contractors love getting the plough out as it is a costly business; farmers hate it for the same reason.) The seed sat there through the drought then grass emerged slowly with a sickly looking fluff, before coming with a rush of the 'compensatory growth' beloved of agronomists when it rained. Now there is a tall sward of lush grass that badly needs grazing to allow it to tiller and thicken in the base. But the soil brought up by the plough has the consistency of wet putty and putting cattle or machinery on it now would take us back to square one.

At least the rain is giving our new gardener Raymond something to do on the grass management front. Raymond, I should explain, is a robot lawnmower. Earlier in the year we invested in a robot hoover, cannily found in the 'middle aisle' of a well-known retailer by the Quartermaster-General. It had been a niggle that the domestic goddess has had to do a lot more mortal domesticity herself than happened in our parents' day, um, pretty well all of it in fact – the rock bottom of a bit of inter-generational jealousy that has gone on since those halcyon Edwardian days when domestic staff on this estate exceeded the entire population we have here now. It seemed only natural, therefore, to name the hoover after my parents' housekeeper Alice; and her counterpart on the lawn after their gardener Raymond. I feel that things are looking up. Hopefully they will soon invent ones that muck out horses, clear gutters, unblock drains, pick raspberries ...

Raymond is apparently the future of farming. No longer will a team of men arrive on noisy tractors, ostentatiously sit having their lunch and leave the packaging in the hedge before tilling the fields. In future, we are told, there will be no need for large machinery. The digital age will be peopled by robots like Raymond busily titivating the soil like an army of ants. I particularly like the idea of the one that will apparently be able to detect weeds with a smart camera and spot spray herbicide. We could have done with something like that this year. The ploughing brought up dormant seed from decades ago so our smart new grass leys are flecked with yellow drifts of charlock, bright white ox-eye daisies and thick clumps of fat hen and redshank. I always marvel at Nature's ability to fight back and something in me thrills at the sight of our fields looking like alpine meadows, bright with chicory, mustard, phacelia, vetches and red clovers. The cows seem to munch it all quite happily and most of the unwanted plants will soon be out-competed by the grass, but we are enjoying the spectacle while it lasts. And our resident bee population, all thirty-seven hives of them, will be busily converting the nectar into honey.

* * *

I can't get the oyster catcher nest out of my mind. Just after drilling I noticed a pair of them were frequenting part of one field near the sea wall. Later I saw the hen sitting motionless, a comical sight; how could a black and white bird with a dayglo orange bill and pink legs think that it could ever blend in on a bare field? And I prayed for the grass to grow quickly to hide her. Then I worked it out. It is the eggs – I found three that matched exactly the clay clods and granite pebbles around them – that provide the camouflage. And at the first hint of danger she would quietly leave the nest and walk in figures of eight around the field pretending with great difficulty to be unconcerned. For two weeks I watched this as crows, rooks and jackdaws fed nearby without finding them. I was just congratulating myself on our new-found ability to put up instant electric fences to protect nests anywhere, off the mains powered dairy fences, when, before I could do so, calamity struck. The heifers escaped from another field, came charging down the track and into that paddock. By the time I got there it was too late. The keening of the bereft oyster catcher parents as they flew back and forth will haunt my conscience for all time.

CHAPTER 10

Into Politics

August 2020

It's very lucky that I am around to write this as a very large limb has just fallen off an old beech tree and gone slap bang across the drive. I had driven past moments before. Someone up there must be looking out for me. The beech is one of my favourites and it now looks rather forlorn with a single stem. The lost limb was the favourite perch of the song thrush that sang so beautifully in the spring that several times we stopped our game of tennis to go and listen. But he will find another stage on which to perform and I'm sure the beech will soon bush up again to compensate. It's the second time it has happened this month, another one did the same last week. I hope there isn't some mystery beech disease going about. It's bad enough to have to contemplate our woods and hedges without the ashes. I expect it's just the wet accelerating the rot in the crevice that beeches always seem to have where branches come off the trunk – a natural design fault if ever there was one.

I'm always superstitious about bad things happening in threes so we are just waiting for the next arboreal upset. At least we have had our three 'getting stuck' incidents now. The first was when I breezily told Davie that the bed of the burn was solid enough for him to track down with the digger to clear a critical drainage ditch. He managed the first twenty yards then had to swim for it as the digger sank into the mud. It took some emergency tree felling and a pull with a mega-digger to extract it before an anxious twenty-four hours changing filters and drying out the engine to see whether it had been knackered. Fortunately, it hadn't. Then a few days later a lorry driver took his satnav's advice (always a mistake in DG2) and tried to make a thirty-ton artic fit through gates made for eighteenth-century carriages, and stuck. This entailed disconnecting him from his trailer, and his breakdown cover sending a crane twelve hours later. Having been very cross with the Polish driver we then felt awful as it emerged that it was his lorry and he had only set up in business the week before. Every drama has a human story attached to it. The third was more prosaic, the familiar challenge of squeezing a sofa made for twenty-first-century bottoms through a door made for svelte Victorian people in one of the cottages. I shall have bruised knuckles for a while yet to remind me of that one. Still, that's the three out of the way and they were really rather minor compared to our worst getting-stuck drama when Bullet the Bull absent-mindedly backed himself into the cattle crush and sat down. It took several litres of olive oil to lubricate him (I got a bit of a row for that) and a pull with ropes tied to the front loader of the tractor to extract him in what seemed like the hardest calving of all time.

* * *

Covid is delaying our on-farm sale that we had planned for mid-August to sell all the machinery no longer required for the new dairy. Quite why a bunch of famers can't follow an auctioneer round a big field bidding for old tractors and trailers is beyond me.

* * *

'Be careful of your acquaintances, as nothing stamps a man more than his friends.' I am gazing across the dining room at my great-grandfather's portrait, where he stands, life-size in his Boer War service dress with a fine cavalry moustache, complacently unaware of the carnage on the Western Front to come. Those words were hastily written in a letter before the First Battle of Ypres, where he was killed shortly afterwards, to be opened ten years later by his small son. I am wondering what he is thinking about the punchy Dundonian sitting opposite me, still wearing his trademark fedora hat. George Galloway has come to lunch and, to my astonishment, to ask me to join him in forming a cross-party alliance he is going to call the Alliance for Unity, with the specific aim of maximising the pro-British vote in next May's election to ensure that the separatists no longer have a majority at Holyrood and a 'government of national unity' can allow Scotland to 'break out of the hamster wheel of neverendum, grudge and grievance'. His analysis is the same as mine, splitting the majority pro-union vote three ways has allowed the SNP and their Green separatist partners to form a government each time on a minority of the vote, and that could go on ad infinitum.

I take to him straight away. He has a natural warmth and a genuine kindness in those pale-blue eyes that I hadn't expected. But it's a big call to make. In the days leading up to his visit I have done some due diligence. He presents television around the world for American and Middle Eastern channels but also, critically, on Russia Today. And that is a concern. On the other hand, that is legal and he is open about that, and front-bench British politicians, including Tories, have appeared on his *Mother of All Talk Shows* on RT. I decide that if the Russians were trying to bring about the break-up of Britain via this election, they certainly would not use a fervent defender of the UK like Galloway to do so. And he is a strong opponent of the cultural Marxism in the 'woke' agenda.

He has also openly supported Irish republicanism, which potentially tars me with an uncomfortable brush. But if anyone has a right to hold that against him it is me. I was blown up by the IRA and lost friends in Northern Ireland. And besides, the hatchets were buried when the Queen shook hands with Martin McGuinness. This struggle in Scotland against separatism is about hearts and minds and if the Celtic-supporting Catholics in the west of Scotland, who form the SNP's core support, are to be won over, then a pro-UK party founded by a half-Irish Catholic may be the best way to do it.

Then there was his unauthorised diplomacy with Saddam Hussein. As a former soldier, who served in the first Gulf War, it's a difficult thing to get my head around, but his initiative was in trying to avert a war. 'Blessed are the peacemakers.' The hardest thing I ever did in the army was to break the news to a widow that her only son had been killed. George succeeded in persuading Saddam Hussein to re-admit Hans Blix and the UN arms inspectors. He is criticised for what he said to Saddam Hussein but diplomacy has its own language and he was hardly likely to insult a brutal dictator in his own palace.

He was also criticised for saying that Iraqis had a right to defend themselves. It is very unfortunate that he used the blunt language he did to reinforce his point but he was arguing the theory of a 'Just War', an important philosophical theory that goes back to St Thomas Aquinas. On balance, many of my generation of army officers now accept that Galloway was right and Blair was wrong, and that George's central point that Baathism was a lesser evil than al-Qaeda and Islamic fundamentalism was proved to be absolutely correct.

Lastly, there were his antics in a catsuit on the *Big Brother* TV show. But it was in a charity event that raised over £200,000. And I like politicians with a sense of humour and the ability not to take themselves too seriously.

George jests that our alliance would be like Churchill and Stalin. And it doesn't take me long to decide that if my great-grandfather's fellow Sandhurst cadet, Winston Churchill, could find common cause with Stalin to defeat the Nazis, then I could – should – team up with George to fight what we both see as the Battle for Britain to avoid the break-up of the United Kingdom. He has an unparalleled record of starting political parties that achieve electoral success. In oratorical terms George would be a very big dog in what promises to be the mother of all dogfights. He has a formidable intellect and I remember him as the most effective debater by far on the anti-separatist side in the 2014 independence referendum.[42] In Effie Deans's words, 'There is something missing where the opposition is supposed to be in Scotland.' And without a serious disruption to the cosy stasis at Holyrood, the best we can hope for is a similar majority for the nationalists and a perpetual continuation of the toxic 'neverendum' that is blighting Scotland.

---

42. It is worth googling George's public speeches from that time and his appearance before the US Senate in 2005. It remains one of the most remarkable performances by any parliamentarian either side of the Atlantic.

If nothing else, a Tory landowner teaming up with Scotland's answer to Che Guevara might demonstrate that there is the very broadest of coalitions to attempt to bring all anti-separatists (the word 'unionist' has unfortunate connotations for many Scottish Catholics like George) across the major parties into one concentration of votes on the ballot to have any chance of winning, something that many have been arguing for. Strange times make for strange political bedfellows after all. Besides, now that the dairy will be run day to day by a manager there will be more time to do other things. I am a great believer in writing different chapters into my life and this promises to be interesting and fun.

We couldn't really be more different and I am uncomfortably aware that I fit all the stereotypes that the other side like to ridicule. When I was serving in Ulster, George was active in the Troops Out movement. Our liaison would be bizarrely eccentric if the cause wasn't so deadly serious. In fact, our intimate knowledge of the Troubles – seen from different sides of the barricades, as it were – leads us both to agree that something must be done to prevent Scotland going the same way. Such Anglophobic hatred has been stirred up by the nationalist movement – organised in the same way as Sinn Féin, so that the SNP is effectively its parliamentary wing – we believe it is only a matter of time before there is a real risk of violence. The Orangemen in Glasgow have been goaded by the divisive Justice Minister, Humza Yousaf's comments after clashes over statues. All it takes is one Burntollet Bridge incident. George, who is a true patriot, though he has perhaps had a funny way of showing it sometimes, is particularly incensed at a recent tweet by an associate of Nicola Sturgeon's encouraging Scots to go down to London and defecate on the Cenotaph. He tells me with pride how he was the only Labour MP to back the erection of a memorial to Bomber Command.

The Second World War is his specialist subject and he has a visceral loathing of fascism in all its forms. I suspect if he had been born a few decades earlier he would have fought with the international brigades in the Spanish Civil War before going on to be a loyal ally of 'Mr Churchill' as George calls him with some reverence. He likes to trace the SNP from its fascist origins in

1934[43] through the war – when its leader Arthur Donaldson was interned as a threat to national security after the SNP attempted to do a deal with the Germans that would have seen him become Scotland's answer to Vidkun Quisling post invasion – to the present day when the *Braveheart* tendency behaves less like Mel Gibson and increasingly more like Hitler's Brown Shirts at rallies and illegal border demonstrations.

We have both been motivated to act by the network of dissident academics and writers, the foremost of whom are people like Effie Deans and Jill Stephenson bravely speaking out against the authoritarianism and endemic corruption of the SNP despite constant intimidation. Whether a second independence referendum is granted or not in the wake of an election victory next year, another five years of the SNP tightening their grip on all Scottish institutions, including, most seriously, the media, could easily take us beyond the point of no return.

August 2020

The clearest indication that we are going through the Scottish Counter-Enlightenment is the absence of one of the Enlightenment's big ideas: common sense, particularly among delivery drivers. For the third time this year a driver from a well-known supermarket's home delivery service has turned down a muddy track and with dogged determination carried on into the ulu, ignoring the branches scratching the sides of his van, the underneath being sandpapered by the hay crop sprouting between the ruts in the centre of the track, and other signs that he – it has always been a he, I'm forced to admit, as the female of the species is more likely to ask the way – is geographically embarrassed. Each time they have grimly stuck to their guns through bottomless puddles until at last they have reached the soft bit just past the sawmill where they have stuck fast in the mud. Except they have not stuck fast. When the Good Samaritan is summoned from writing a *Country Life* article

---

43. The SNP were the Scottish equivalent of Mosley's Blackshirts. Their logo today is a more rounded version of its original symbol, the odal rune, a letter from an ancient Germanic alphabet, which was also adopted by 7th SS Volunteer Mountain Division.

to help pull them out it is evident from the tyre marks that they have battled tenaciously with the mud, like a fox in a snare, and dug themselves in more deeply. So there has then been a *Krypton Factor* challenge with ropes and chains, sometimes by torchlight. Bringing home the bacon has never been so exacting.

We tried putting a chain across the track but to no avail. Nothing, it seems, will deter a driver who has been programmed to listen to his dashboard. I think the only answer will be a large road sign on the edge of the village that says, *You are now leaving civilisation. Please switch off your satnav and ring for directions if unsure.* And as for the clever clogs in Silicon Valley who thought it was funny to centre our cyber-postcode on the muddy dead end of a cart track, an especially hot corner of hell awaits.

* * *

This hot, thundery weather has been unsettling for man and beast alike. I have noticed bees threatening to swarm and the hens have been very unreliable in their laying. I daresay it has also helped to bring out my inner Basil Fawlty. I hope our holiday cottage guests haven't noticed. Certainly they have all been very appreciative of the chance to escape lockdown and enjoy the delights of the Costa del Solway. All, that is, except the couple who arrived on a Saturday and waited until it started raining on Friday afternoon to announce that they were leaving and would expect a refund, or they would post a bad review on a global online booking agency's website. I console myself with the thought that the review will probably deter anyone else with third-degree OCD from booking with us in future. The beaches look complete again with clusters of children around rock pools like the shell seekers in a Dame Laura Knight painting. And on my rounds I am heartened by the waves from happy punters enjoying walking around the estate.

The fields are looking bonny. The changed weather has saved the droughted crops from what seemed like disaster a few months ago, and harvest has so far been bountiful. Nature has a way of snatching solvency from the jaws of bankruptcy. The new grass takes on shades of shocking green when viewed through moist air and is putting a pleasing bloom on the heifers, now six months away from calving and our first pints of milk. In the meantime we are anxiously tracking the parlour on the high seas on its way from New Zealand.

* * *

The madness of lockdown has produced many disturbing side effects but none more so than the narrative being pushed by the anti-Boris chorus in the media – that of Nicola Sturgeon's competent leadership during the pandemic. This is the modern equivalent of the 1930s trope about Mussolini making the trains run on time. I use the comparison deliberately. The North London intellectuals applauding Sturgeon for her anti-Brexit stance are deluded. The first response of the Sturgeon administration to Covid-19 was to attempt to suspend trial by jury. Prior to that they had fought tooth and nail to bring in culturally Marxist legislation to appoint a 'state guardian' for every child in Scotland to oversee how parents bring up their children. Their main effort during lockdown, aside from spinning the factually unsupported story that Sturgeon's carefully differentiated approach to tackling Covid has been superior, has been an attempt to crack down on free speech via a scarily open-ended Hate Crimes Bill. Sturgeon and her separatist friends should be anathema to the BBC-*Guardian* intelligentsia.

Despite all this, opinion polls are showing a majority for separation and the bookmakers are currently quoting the SNP at 5/6 on to form a majority government in Holyrood next May. The electoral registration office in the council sends us the usual form to check our ability to vote. The envelope says that fourteen-year-olds can now be added to the electoral roll. There is only one possible deduction one can draw from the SNP's tinkering with the franchise, which includes allowing those with no fixed abode a vote. The 2021 election will be a dirty fight. Some Scottish opposition parties are burying their heads in the sand and their attempts to portray the election as 'not about independence' and 'focusing on the SNP's domestic failures' only reinforce the sense of the impending partition of Britain. I despair of the utterly abject appeasement of Sturgeon by the opposition at Holyrood since 2016. To needle them into being more robust I start using the hashtag #Appeasition on Twitter.

George and I are pounding the airwaves on YouTube (conventional TV has yet to engage with us[44]) to put pressure on the unionist parties not to stand against the strongest pro-union candidate in the first-past-the-post constituency ballots while we make best use of the d'Hondt AV system we have in Scotland by running pro-union Alliance candidates from across the political spectrum in the ballot for the list seats. If our candidates are elected we resolve to sit as independents to help to form a government of national unity in Holyrood that will start to restore neighbourliness across the UK and particularly in Scotland.

In the meantime we need candidates. The first ones to rally to our colours are all disaffected Tories like me. I tell George that I am relying on him to find me some raving Trotskyites to balance things up. He replies, 'Steady on Jamie, I don't think we need go that far!' Actually I am beginning to understand that George's Workers Party (strapline 'for the workers not the wokers') is really socially conservative, if well to the left of me economically.

An early supporter is the former Scottish rugby international Finlay Calder. By strange coincidence he tells me that he used to buy wheat from my father during his second career as a grain trader. Finlay is a tower of strength. He rings regularly to give me motivational talks, explaining how as captain of the British Lions he motivated the team to lead them to victory in the 1989 tour of Australia. He sends me a video of *The Gathering Storm* and tells me to watch it.

Our enemies are seizing on George's and my seeming lack of cohesion as proof that the whole project is doomed to failure. But it is in fact a strength as it demonstrates the breadth of the grassroots movement building momentum to take on the nationalists. There is no requirement to defend each other as we can rightly assert that we are combating the groupthink that has bedevilled politics by fielding independently minded candidates who will bring experience from all walks of life, to leaven the brash young people in suits who sit there at the moment, by sitting in the parliament as independents. Our policy is that all A4U candidates will be free to vote according to their individual consciences as MSPs – democracy closer to its Athenian origins for the Athens of the North.

---

44. And hardly did throughout the campaign.

All new political parties say that they are introducing a new type of politics, but I think we must be the first party in history to promise in effect to dissolve our party as soon as we are elected. The early signs are that people like what they are hearing. After three weeks we have many more followers on Twitter than the Scottish Liberal Democrats and momentum is continuing to build. The challenge is probably not in persuading the electorate that Scotland deserves better than the separatists' authoritarian one-party state, but in persuading the older unionist parties that the pro-union parties really are 'better together' in what promises to be a bitter fight.

* * *

My faith in humanity is restored by the kind man from SOS Drains who climbed into one of our septic tanks to retrieve my drainage rods. If there is one feeling worse than that of losing a salmon it is the sudden loss of pressure as the rods part company, leaving the end with the plunger on halfway down the pipe. Fortunately the jetter pushed it down into the tank. I was beating myself up for my incompetence when I saw that the offending rod had actually broken where the metal thread is joined at the end, so I had the satisfaction of being able to blame my tools instead.

* * *

I am three weeks into my new role as Deputy Leader of the Alliance for Unity and the pattern of my life between now and the Scottish Parliament elections in May is starting to clarify. I have just been speaking to a man called Goat, who might make a YouTube documentary showing me travelling around with George, highlighting misgovernment and unhappiness in Sturgeon's Scotland. Apparently the 'odd couple format' is regarded as 'an evergreen' by the men behind the lens. George is much better educated than I am and calls our partnership the most unlikely political alliance since the Fox-North coalition.[45] When I have

---

45. The coalition government of 1783. Fox (an Old Etonian!) is a radical hero of George's.

looked it up I point out that Charles James Fox and Lord North didn't actually stay in business together for very long.

In fact we agree about more than we disagree on, particularly the shocking authoritarianism of the SNP; though perhaps it is fortunate that defence and foreign policy, on which George often has controversial views, are matters reserved to Westminster and therefore don't need to be discussed. If, by some miracle we both make it to Holyrood he will no doubt be arguing to put taxes up while I will be wanting to lower them and we quite like the idea of being 'frenemies' if we are successful. In the meantime we are enjoying hamming up our relationship. He tweets that he has just had lunch with his favourite living Old Etonian and when the usual suspects start to throw cyber-bricks he points out that Tam Dalyell and George Orwell were also schooled by the Thames.[46] I have the strange sensation of being re-tweeted regularly by lefties of the Workers Party now, so we must be breaking down barriers.

The other recurring theme of my daily conversations is defending my decision to incredulous friends. There has been less opposition than I had anticipated as they all agree that the opposition to Sturgeon from the Scottish branch offices of the Tory, Labour and Liberal parties has been risible. What exercises them, and what exercised me initially, is the fear that we might 'split the vote'. The logic that the vote is already split three ways, and so we are doomed to failure and another half decade of the SNP anyway, or worse, unless we do something radical, takes a while to sink in.

Being deputy leader of a political party, even one as nascent as ours, implies being surrounded by a host of spin doctors and leggy interns making coffee and running about with clipboards. Nothing could be further than the truth and the headquarters team is as lean and dispersed as a terrorist cell. Although we have a growing candidates list. We are keen to avoid identikit politicians and thrusting PPE graduates so we are deliberately trying to find people who have 'done stuff'. It includes ex-Regimental Sergeant

---

46. George tells me that he has spoken at Eton several times. He took the Eton chapter of the Stop The War Coalition to a rally in Hyde Park and introduced them on the stage to great cheers. He wonders where they are now and we agree that they are probably investment bankers.

Major Arthur Keith of the Black Watch. I enjoyed ringing Arthur to congratulate him with the traditional army greeting, 'Stand by your beds!' and we had a good chat about the similarities between the SNP and their soulmates in Sinn Féin, with whom Artie and I have had dealings in the past.

One of the first things on our to-do list is to register the party before the SNP dark arts department spike our guns by registering one in our name. Happily we get our application in to the Electoral Commission first, although it is clear that the separatists take us seriously. Stunned by our rapid acquisition of followers on Twitter they set up a parody account in no time at all. And there has been a steady onslaught of online bullying, fake news and smears. Very post-modernist of them.

* * *

There is an air of defeatism in the unionist camp. There seems to be a certain type of columnist, either right wing or on the Left, who cannot forgive the present Tory government for Brexit and likes to use Nicola Sturgeon as a rod with which to beat Boris. It undermines the union and blinds them to the SNP's deplorable record in office and extreme intolerance.

Morale is boosted by Scotland's leading pro-union blogger Effie Deans coming out in favour of Alliance for Unity in her blog. She describes the Scottish Tories, Labour and Liberals as having as much chance of defeating the SNP on their own as Cowdenbeath FC have of beating Real Madrid. She calls on them to work together with the Alliance for Unity to translate the silent, anti-nationalist majority into a majority of seats at Holyrood. #CowdenbeathParties starts to gain traction.

* * *

September 2020
My eye is caught by a thistle head swaying and I strain my eyes to see what has stirred it. As my pupils focus I see a whole charm of goldfinches cleverly camouflaged in the weeds. It never ceases to amaze me how a yellow, red, black, white and brown bird

sitting on a purple flower can be so hard to see against a green background. At least there are a few thistles – in a paddock sold to some neighbours many years ago; and their laissez-faire attitude to grassland management helps to boost the food supply for the finches, helping to salve my conscience a little.

The fields and hedges are dotted with family groups of birds at the moment. Three times I have had to stop the pick-up to allow hen pheasants with small broods of poults to shimmy down the road before fluttering into the hedge. And the brood of wagtails whose nest we anxiously watched through the summer are now to be seen roosting along the ridge of our roof at dusk. Down on the shore parties of waders are out feeding on the mud, the avian equivalent of families pulling in to Tebay Services for a burger and chips on their way back to their autumn term-time quarters.

The suckler herd is weighing heavily on my conscience as I gear up to sell the last of them. There are few better sights than cows and calves in late summer. The knowledge that this is the last year I will be able to enjoy them is sad. I have bonded with the new dairy heifers and like watching them but dairy cows never quite seem to achieve that characterful matronly look. Some of the sucklers are old friends who have been in the herd for fifteen years. One consolation is that dairy cattle are at least safer. Beef cattle have never reached that level of domestication and though, touch wood, we have always had quieter cattle than most, there has been the odd one that has been a bit 'fast'. And Davie and I have started to worry that as we get older we might not be able to get out of the way quickly enough.

***

I am avoiding looking out of the window at yet another 'soft day' when a perpetual drizzle reinforced by short sharp showers soaks man and beast alike. And I am trying hard not to think about the last field of barley going flat. At least it is reinforcing my decision to put more fields down to grass for the dairy and reduce our exposure to arable cropping. The pigeons are having a lovely time feeding on the spilt grain.

***

The opinion polls are bad again. Brexit and Covid may have been traumatic episodes but their lasting effect may be existential to Great Britain if they lead to the loss of Scotland and the break-up of the union. English nationalists who dream of ridding the union of Scotland have not thought things through. A weakened Britain, no longer technically 'Great',[47] would be a very sorry realm and a tragic closing chapter to the reign of Queen Elizabeth II. Unable to defend the northern approaches, or hide nuclear submarines in sea lochs, no longer an oil producer, shorn of over a third of its landmass and probably with a government that had lost all authority if not actual power at one of the most vulnerable times in our history, there would be little for Little Englanders to celebrate. An impoverished Scotland might easily be driven into the arms of Russia or China.

Westminster is sticking to the line that the 2014 referendum was a once in a generation event but even staunch unionists concede that may be untenable in the face of persistent non-cooperation from Holyrood and carefully orchestrated civil disobedience leading to the Ulsterisation of Scotland. And it is tempting to think that it might be preferable to risk a second and final vote – as happened in Quebec – than have another four years of hate-filled 'neverendum'. If only Tony Blair had listened to Tam Dalyell's warning about devolution being a 'motorway, without exits, to independence'.

The post-truth fake history deployed by nationalists with the help of Mel Gibson has somehow embedded the notion that the union was a one-sided affair resulting from subordination of Scotland as some sort of colony. The truth of course was that the Stuarts were primarily Scottish, even if they subsequently preferred the pleasures of Newmarket. And possibly because, unlike the unions with Wales and Ireland, this was a genuine marriage of equals when the parliaments of England and Scotland voted themselves out of existence and the two kingdoms merged into one, most other institutions, the church, the judiciary and, later, football teams, were kept separate and the idea of two

---

47. The name Great Britain was formally adopted in Article I of the 1707 Act of Union.

separate nations was perpetuated in a way that it was not in say, Prussia and Hanover. That has lent the illusion of legitimacy to the decentralising dynamic.

Another unfortunate deficiency has been the lack of any provisions for secession. It never occurred to anyone in 1707 that the union would ever be severed democratically. The threat at the time was from Jacobite invasion. The legislators themselves, but by no means everyone, probably also didn't want to contemplate the union ever ending; it had been asked for partly to end a bitter trade war in which an English ship had been seized by the Scottish authorities and the crew hanged on a trumped up charge of piracy. They did not want to invite any return to hostility.

We pride ourselves on having an unwritten constitution but sometimes these things are better written down for the avoidance of doubt. As we now all know, the EU did not allow this omission, possibly that strengthened their union and made it harder to leave but certainly it meant that the Brexit debate was based on known unknowns. By contrast the Scottish separatist debate is largely based on unknown unknowns because the rules of secession have never been codified. The Canadians had the same issue and, following the Quebecois referendums, passed a Clarity Act in 2000 to establish the rules for any future secession referendum. One of my Twitter followers named Peter draws my attention to this and the more I think about it, the more convinced I am that the Alliance for Unity should press for a UK Clarity Act and I devote myself to working up what it might involve.

Such a move would no doubt provoke howls of faux constitutional outrage from Holyrood – although the SNP 'government' is quite happy to pass legislation itself on this issue – but the fact remains that the UK Parliament with its fifty-nine Scottish MPs remains sovereign and is the only place where this legislation can legitimately be executed. After discussion with George, and the panel of academics and business people I have assembled to guide Alliance for Unity policy behind the scenes, we come up with nine questions that need to be answered.

First there must be a defined procedure for requesting a referendum. What criteria must be met? Nicola Sturgeon claims a 'mandate' for indyref2 on the basis of a majority of seats but there has never yet been a majority of the popular vote for separatist parties in an election. That mandate must be defined.

Secondly the franchise and by extension citizenship. Prior to the 2014 vote, concessions were made to the separatist side, who had lobbied for the voting age to be sixteen despite remaining eighteen for general elections. The SNP would give the vote to a sixteen-year-old criminal, born overseas but residing in a young offenders' institute; but the captain of the Scottish football team, who invariably plays for an English Premier League club, would be disenfranchised. The ad hoc nature of the legislation also meant that Scottish servicemen serving outside Scotland were denied a vote on a referendum that would have turned them into foreign mercenaries in the British armed forces. Most seriously the Scottish diaspora spread across the UK was denied a vote in a referendum that might have deprived them of British citizenship. Since Article XXV of the Act of Union repealed the English Alien Act that treated Scots as, well, aliens, there has been freedom of movement across the UK and a shared citizenship. Like nearly every generation of our family before them, my children are enjoying their young adult life in London hopefully before returning home to Scotland, as I did. Current proposals for any independence referendum would deny them a vote on their futures. These inconsistencies must surely at least be properly addressed in parliament rather than haggled over in the run-up to a vote.

Thirdly the wording of the question should reflect the seriousness of the decision. The Cameron government believed that by giving Salmond every advantage, including the Yes side of the argument, if, as they confidently believed, the answer was still No, the victory would be all the more convincing. Unfortunately it hasn't worked out that way and asking Scots, Yes or No, whether they think Scotland should be an independent country has created an unfortunate precedent. Asking Scots whether they wish to Leave or Remain in the United Kingdom would be a more accurate and appropriate question.

Fourthly there should be clearly defined minimum requirements for the threshold and the turnout. The 50 per cent threshold set in 2014 was historically low compared to similar referendums around the world.[48] Had there been a very small margin for independence on a low turnout it might have resulted in a majority of Scots being bitterly resentful of being deprived of their British passports.

Much would depend on whether it was planned as a one-phase or two-phase vote and this is linked to the fifth question over whether Scotland should be treated as one homogeneous entity – as it was at the time of the union – or allowed to vote regionally. A second confirmatory vote would give unionist Orkney, Shetland, Dumfries and Galloway and Borders, and perhaps other areas, all of which are now actively discussing separation from Scotland, the opportunity to opt back into the United Kingdom. The alternative could be plunging a newly independent Scotland into a secessionist nightmare as the former kingdoms and principalities protest against 'being dragged out of the union against their will'. There is a democratic deficit in Scotland: rural Scotland seldom gets the government it votes for as the population is skewed towards the urban central belt. It would be better to give our citizens a chance to rectify that at the time of the vote.

A phased vote might also help Scots decide their futures based on known knowns rather than the current unknown unknowns. Questions six, seven and eight relate to debt, pensions and trade. SNP Scexit planning currently makes the blithe assumption that Scotland would be allowed simply to walk away without any share of the national debt, lending new meaning to the expression 'Scot free'. Parliament needs to express its decision clearly on how, or less likely if, the UK national debt would be apportioned before Scotland could be recognised as an independent nation with powers to borrow herself. Linked to that is the question of pensions. Scotland would be unlikely to be able to fund the present level of pensions, the alternative is for the former UK[49] to pick up the bill but is that really tenable? Similar assumptions are made about trade and the question of a hard border. There was one prior to 1707: would there be again? We don't know and the mechanisms for agreeing future trading arrangements must be clearly laid down.

---

48. And indeed for changes to the SNP's own constitution, which requires 60 per cent.

49. I refer to former UK (fUK) rather than rest of UK (rUK), because it would be.

Allied to this is the question of currency but that seems to have been settled, at least as far as England is concerned: the Scottish treasury would have to make its own arrangements.

Finally, question nine relates to the provision for any future vote in the event of a decision to remain in the union. Clearer rules on the minimum elapsed time before another referendum and the conditions to be met would avoid a repeat of the bitter 'neverendum' we are enduring.

When we review all our ideas George and I decide that this is the silver bullet that can defeat separatism in Scotland once and for all and decide to put the Clarity Act centre stage in our manifesto.

The Critic publishes the Clarity Act proposal as an article and *The Times* takes a shortened version in the Thunderer column. When I introduce the idea on Twitter with the hashtag #IndyPostcodeLottery it drives the nationalist trolls berserk, a sure sign that it has found a chink in the SNP's armour.

CHAPTER 11

Farewell to the Beef Herd

The stillness of a late September day is seasonless, neither hot nor cold, dry but with a hint of damp. Just plain vanilla weather. There is still the odd swallow but, every now and then, I hear curlews calling on the foreshore and geese honking overhead, small reconnaissance patrols feeling their way south from the Arctic Circle. I have just seen a wheatear flitting between the stobs along a cow track. And one morning a pair of egrets fed alongside the cattle as if they were beside water buffalo in the Nile valley. Birds are on the move. Yet it's hard to be enthusiastic about the countryside right now; neither summer nor fully autumn, the leaves are tired without showing much colour and the land grows grass grudgingly or sits in sullen stubbles. The triumphs and disasters of the growing season are behind us. We have taken to walking the dogs at sunset, trying to squeeze the last little bit of light out of the day. It is often the best time to see wildlife. Sometimes we ambush the roe deer and they bounce away into the gloom, their tail lights glowing white long after the rest of them has been swallowed by the darkness. Coming back along the beach we are startled by a strangled cry and turn to see the old heron lifting out of the shallows and heaving himself up into the tops of the trees to roost with his strange dinosaur flight.

October 2020
The old cock grouse turns and comes skimming downwind in front of the line of butts, contouring the ground with magnetic precision. Instinctively the stock of my gun finds its way to my shoulder and

hugs my cheek like a pillow as the tip of the muzzle swings round below and ahead of the bird and I fire. The grouse crumples and cartwheels into the heather, the neat dark missile transformed in an instant into a duster of brown and white feathers. And I bask in the glorious sunlight of an early October morning. Somewhere in the distance a grouse calls *go-back, go-back,* and then I wake with a start, it's not a grouse calling, it's someone wanting to speak to me. That can only be bad news early on a Sunday; probably someone has arrived in a holiday cottage last night and is annoyed that the Wi-Fi is not as fast as in their Edinburgh flat, or the heating thermostat is not set as high. Or maybe we have stock out. It's the latter: the voice at the other end has a brittle timbre. 'They've been right through the wood into the garden and all across my beautiful grass.' I mumble apologies and curse the cows as I pause to savour the warmth inside the bed before stumbling out and into my clothes. Beside me the Assistant Cowherd murmurs a comatose 'I'll come and help' without opening her eyes. The first lie-in for weeks ruined.

When we arrive at the crime scene the cows lift their heads and give an insouciant, sidelong glance before casually turning with deliberate slowness and picking their way back through the trees to the gap in the fence. I reflect on the predisposition of cattle to invade woodland at any opportunity. They benefit from grazing the different herbs, and the habitat of the forest floor is improved by the ground being poached and manured by the descendants of the mighty aurochs of the primordial wild wood. I daydream briefly of woodcock poking their beaks into the hoof slots when they arrive from Scandinavia next month. Dutifully the cows file through the twisted and flattened fence where the stobs have snapped (I blame the EU's ban on proper creosote). All except one truculent bull calf who disdainfully turns to where the fence is still standing and jumps it off his hocks with a cervine ease, as if to let us know that he doesn't think much of it. It's the last act of rebellion of our beef suckler herd.

Tomorrow the last of them go and we will be dairy farmers again, except on the fields too far from the parlour where we are continuing to grow cereals and potatoes. This strange non-event of a year will forever be associated with their departure as 2020, 'the year we put the beef cows off.' Though no doubt historians will remember it primarily for other reasons.

At least the farming side of the business has continued without interruption and our labour has been rewarded when our produce has been sold. Not so the other rural businesses who have worked through lockdowns to keep their land from reverting to scrub, the groundsmen on the golf courses who tended the greens and fairways so that they were in mint condition when we were *finally* allowed to play again, the ghillies who kept the river banks mown for absent fishermen and the gardeners who toiled heroically in isolation only to find that, 'Full many a flower is born to blush unseen and waste its sweetness on the desert air.'

* * *

The sense of limbo is particularly acute this year as we transition from beef to dairy. For a brief few months I don't feel compelled to check the price of beef or milk in the papers, much like a recovering gambler must feel at not following the racing form.

The farming lull is enjoyable. There is no tearing hurry to plant or harvest crops. But the paper war goes on ... and on. I am just congratulating myself on selling the last of the sucklers when a gnawing feeling tells me something is wrong. The list of numbers from the passports doesn't tally with what is written in the ear tags. The cow in question, an Angus cross first calver looks blissfully unaware as she flutters her eyelashes at me under a mass of black curly locks. It must have been the hair obscuring her ears that led us to mistake 201380 for 201386. It wouldn't matter a jot in almost any country in the world but here all bureaucratic hell will break loose if I don't resolve it, which will be embarrassing as her partner in this 'crime', as the agents of the Green State may well call it, went in the last consignment and is now grazing innocently on another farm. There had to be a final humiliation in my cattle record keeping.

Then the ongoing saga of the Scottish Executive's Agri-Environment Climate Scheme. We have been in the scheme five years yet the government computer has yet to be completely happy about it. It has spat out my annual claim form. Today's mail contains a 'Dear

Producer' letter telling me that I must 'submit an environmental assessment if you state you are undertaking alternative practices or mixture of production and alternative practices. This should have been submitted by 31 August 2020.' I hold my head in my hands thinking about all the environmental assessing and form filling that has accompanied this (expletive deleted) scheme since its inception, the money that has never arrived, the extra man-hours expended, the endless inspections to measure field margins and hedges, and issue a silent scream.

When the lady from the department rings to assure me that this is an error and I don't have to do anything, she expresses concern that I haven't applied to extend the scheme. I politely tell her we have decided it isn't for us after all. The environment and the climate are important but nothing is more critical than my sanity.

***

When the history of the last decade is written – when I wrote 'last' I immediately thought that it has really been a 'lost' decade when Scotland has been stuck in reverse – the role played by the Scottish Green Party will come in for special criticism. I had never really thought much about the Green Party. While serving around the world in the army I had formed a hazy impression of them as well-meaning eco-eccentrics, the recipients of the none-of-the-above vote at elections and recently led by Caroline Lucas, who seemed to punch above her weight and occasionally say something sensible on *Question Time*. It was only when I moved back to Scotland that I realised that the *Scottish* Greens were something entirely different, and only very tenuously connected to their English counterparts. In fact it took an old friend of mine who has stood with repeated and so far unfulfilled optimism as a Green Party candidate in the south of England to explain the difference. 'You see,' he said, 'you and I are limes. We care passionately about conserving the environment and we are green on the outside and green all the way through. That lot in Scotland are watermelons, green on the skin and red to the core.'

That definition helps to understand everything about 'the SNP's gardening section' as George likes to call them; it explains their heady mixture of land reform agenda and romantic back to nature ideas of returning Scotland to its pre-agrarian state. It is a deeply misanthropic philosophy. The Greens would like to see shepherds driven off the hills along with their sheep and replaced with forests replete with wolves.

The SNP and the Greens became natural bedfellows as the SNP are also misanthropic, unless you are a racially pure Scot and support independence that is. The Greens, for opportunistic reasons, also supported separatism and the SNP were delighted about this as they were both then able to game the electoral system with the SNP focusing on the constituency seats and the Greens winning seats off the List in an unofficial separatist alliance. Then lo, in 2016 the Scottish Greens found themselves kingmakers and propping up the minority SNP government.

Since then the gardening section has greedily extracted concessions from the SNP in return for supporting their legislation, and in doing so has forced it to give up any pretence of being 'Tartan Tories'. And the Scottish countryside has been battered every time the SNP has had to throw their confederates some red meat in order to get legislation through. Most recently, the ban on controlling mountain hare numbers threatens the fragile ecology of our heather moorland.

During all this time Scotland's countryside has suffered. Wild salmon have disappeared from our rivers while poorly regulated salmon farms have spread disease. The red squirrel has become extinct across much of Scotland as the pox-bearing grey squirrel has spread. Wading birds like the curlew have also declined rapidly as the heather moorland where they breed has been covered in trees, and badgers and other predators have proliferated unchecked. Deer have been systematically massacred and left to rot on the hill to make way for yet more trees. Yet the 'gardening section' has turned a blind eye to these problems and instead stoked the class war against fox hunting and grouse shooting and driven a weirdly woke agenda for Scotland's children.

As a farmer I have watched with deepening gloom as this unholy separatist alliance has ridden roughshod over rural interests, ignoring the advice given to them by organisations like the NFU and the Scottish Gamekeepers' Association. And my heart soars when I hear George Galloway say, in a YouTube broadcast, that the Alliance for Unity will 'obliterate the gardening section at the polls'.

November 2020

A celebrity Twitteress tweets, 'According to my nipples, summer is over.' I can't say I've noticed; maybe blokes don't have the same seasonal indicators, but it is certainly getting a lot colder. Normally I get depressed at this time of year as we fend off the winter regime of feeding and bedding housed cattle and dealing with mud and muck and anxiously eke out the last of the grazing in the drier fields. But this year I can barely contain my excitement. The new dairy rises like a temple to a sacred bovine deity and the anxieties are now whether it will be on time and fit for purpose. The law of averages states that we will be driven off the land by the weather and need to house cattle during November and we need the milking parlour ready for when the first calves arrive at the end of January.

Going from straw courts to individual cow cubicles, complete with rubber mats, is a bit like demolishing 1960s council tower blocks and building smart new houses. The sheds are dormitories flanked by feed yards where they will wander out at will to eat silage from large feed bunkers. Where previously suckler cows lay in a deep litter all winter with 'tackles' of matted muck and hair hanging from their coats, so that we worried about the calves mistaking them for teats when they arrived, now they will be clean and the muck will be pushed down a passage and across slats into a large tank then gravity fed into a football field sized lagoon. The hassle of bedding the pens, the dust and the straw being blown around the yard, are becoming memories on the continuum into the past back towards my earliest experiences of seeing cows being chained in dark byres all winter. The lagoon has been expensively lined with a thick membrane and then covered with another reassuringly expensive plastic sheet so that greenhouse gases and smells are kept contained. The whole site is sloped at ten degrees into the slatted tank so that dirty water from any surface touched by a cow ends up in the lagoon. Every available particle will be saved for valuable slurry to be pumped back onto the land to maintain our fertility, while rainwater from the roofs has its own network of underground pipes to the ditch along with any rain that falls on the lagoon cover and is pumped off. Solar panels on the roofs will power the whole enterprise during the hours of daylight. When the green Stasi come to audit us I am hoping to be able to write that the whole enterprise will have a microscopic, perhaps even a 'zero' carbon footprint when the voracious consumption of carbon dioxide by all our lush grass is calculated. Then I can argue against the Neo-Roos from the moral high ground.

The whole enterprise has been planned with 'cow flow' in mind so that all the cows' energies are concentrated on milk production rather than moving from A to B. The dairy sits in the middle of the farm and a network of tracks lead to the collecting yard at the top of the parlour. Electric fences, bungees, gates and pens will help to divert them where we want them. It is very satisfying seeing it all take shape but I can't wait for it to end and for the noise and bustle of diggers clanking and men yelling to disappear so that I can start the most enjoyable phase of all, making it all sit beautifully in the landscape and restoring as much habitat as possible to the patches of waste ground left by our new 'industrial' complex. I have visions of birds nesting in thick patches of dogwoods, thorns, hazel, holly, dog rose and gorse around the lagoon and the silage pits, and a screen of trees: willows, cherries, rowans, whitebeams, hornbeams and crab apples breaking up the outline of the buildings. And the new quarry where we have dug all the stone for the development will be turned into a pond in a wood and be springing with teal. A winter of hedge and tree planting awaits.

***

As a tribal Tory on secondment to the Alliance for Unity I have met with some hostility on Twitter from dyed-in-the-wool Tories, particularly from career politicians who perceive that we sit between them and the Holyrood trough. I had never seen their sense of entitlement from this angle before and it's very revealing. But events last week have hardened my resolve to work with George to try to convince the anti-nationalist side of the electorate that, if they want to avoid another four years of 'neverendum' or worse, they must persuade their tribal leaders to change their approach and unite.

Last week should have been a disaster for Nicola Sturgeon. She seemed close to resignation after inconsistencies in her story about the Salmond Inquiry appeared to suggest that she may have misled parliament and allegations that her husband may have attempted to influence the course of justice. Margaret Ferrier was perhaps the most despised person in Scotland after knowingly travelling the length of Britain by train with Covid before taking the virus into the House of Commons. Sturgeon was closing pubs in Glasgow in the run-up to the Old Firm match,[50] something not even Hitler achieved in the Blitz. Yet by the end of it an IPSOS MORI poll had showed a 58 per cent support for the partition of Britain, the SNP's twenty-something-year-old Finance Minister Kate Forbes had knocked Douglas Ross, the leader of the Scottish Tories, around the ring on *Question Time* and the SNP had actually won a council by-election in a Tory heartland in Aberdeenshire by splitting the anti-nationalist majority vote.

It's going to be an uphill struggle between now and May, particularly as the Tories, Labour and Liberals only seem interested in battling it out between themselves to see who comes second, third and fourth.

* * *

The human cost of the lockdown is starting to be counted in metaphysical ways. How many love affairs have been nipped in the bud? Brief encounters condemned to remain forever unrequited by the carelessness of a Chinese laboratory technician. How many Mrs Bennetts have been thwarted by the absence of Mr Darcys in the neighbourhood this year? But in our household the heaviest amorous sacrifice has been canine. The year 2020 was meant to be when Pippin, our Norfolk cross Lucas (Norcas?) terrier was to find a husband. We were thinking about finding a suitable beau for her to be 'put to', in hunt kennels parlance, when the world ground to a halt. We resort to shameless free advertising in my *Country Life*

---

50. Glasgow Rangers v Celtic.

column in the hope that it will yield some good offers of marriage. 'Only dogs with the finest pedigrees, preferably blended with judicious hybrid vigour, and GSOH will be considered. She prefers a bit of rough so smooth-coated types need not apply.' The social drought of Covid is enlivened by love letters from terriers all over the UK.

* * *

The lowering November skies are lit by golden candles of larch on the hills, now at the peak of their brilliance. It has been an autumn of change on the cattle front. It seems, in equine terms, as though we have gone from Clydesdales to Arabs. And the change in scale is taking a bit of getting used to. The fields should really be empty, now that it has turned wet, but the new sheds are not quite complete so the cattle are moved between sandy hillocks and rushy bogs to save the best fields from damage, much to the delight of the snipe that have arrived in force to feast on the worms in the mucky soil. It is a blessing that the heifers are as light as they are, probably only half the weight of some of our old suckler cows, so the compaction is not as bad as it might have been. Even so, it is an anxious wait while the builders lay the last of the concrete. The unrelenting wet means that some of the heavier fields will not be able to be grazed until the spring – not by cattle anyway. The large black and white locusts known as barnacle geese have spotted the opportunity and the pastures are soon littered with their droppings and feathers and the air filled with their chatter.

* * *

It's the season of remembrance and the kitchen is red with poppy paraphernalia as the hub of the parish's collecting campaign. This year is especially challenging as Covid has meant that the pubs and shops where we usually place tins have closed, so door to door collection will be critical. I wonder whether our team of volunteers will still be prepared to do it but they are made of stern stuff and immediately suggest pragmatic ways of socially distanced collecting. The edict from the Scottish government that outdoor

Remembrance Services at village war memorials are not to take place has met with similarly stout resistance, I'm pleased to say, and will be given the Nelsonian eye. The pusillanimous directive that we should stand on our doorsteps to observe two minutes' silence instead has gone down particularly badly with those of us who are, in modern terminology, 'veterans'. I can't help thinking that this is another SNP attempt to undermine the concept of Britishness. The state's attempt to impose itself on our annual tribute to the fallen, for many of us a very personal reflection on the friends we lost, is deeply shocking. In the depths of the countryside where few can see anyone else's door it would be a futile gesture anyhow. The young men whose names are on our memorial, my great-grandfather among them, must be wondering what freedom they were fighting for. It gives an added impetus to make it an extra-special service this year, and what Edmund Burke would have recognised as 'the little platoon' swings into action. There is a setback when John, our regular piper, tells me that his arthritis has forced him to give up playing now that his fingers can no longer navigate the chanter. A call goes out to a promising schoolboy piper in the town, so that 'Flowers of the Forest' can fill the Kirkbean Glen with its haunting lament again.

We will remember them.

* * *

The Alliance for Unity has come a long way in a few months. I am humbled by donations from friends and neighbours, including fellow cadets at Sandhurst I haven't seen for decades, but we are desperately short of the money we are going to need for leaflets, billboards and advertising if we are going to cut through. Part of the resistance from would-be donors seems to be anxiety about George. In fifty years in politics he has some inevitable baggage and he is the first to admit that he is 'Marmite' to some. He suggests that I should take over as leader of the party while he continues to spearhead the campaign as 'lead candidate'. It takes me by surprise but I can see the sense of it and agree to do it, with the proviso that I would stand aside if we can attract a big household name like Neil Oliver to do it. It will at least help to diffuse the charge that this is

George's vanity project. He is genuinely only interested in the cause of defeating nationalism and is actually sacrificing quite a lot of his time that could be spent very lucratively on his broadcasting and writing. We tell the candidates in a Zoom call and I explain that there has been no coup and, in reference to Leon Trotsky, George is not walking about with an ice pick in the back of his head. George growls that my historical knowledge of Bolshevism is rubbish as he would have been the one wielding the axe not receiving it.

I decide that as leader I need to grapple with A4U's vision. We have our Clarity Act idea but there needs to be more for our candidates' to unite behind. Our call to action struck a chord with many people from across the political divide, angry at the direction that Sturgeon's sub-fascist nationalism was taking Scotland, seemingly with an impotent opposition unable to do anything about it. George's simple message resonated: the Scottish Conservatives, Labour and Liberals trying the same failed strategy at every election with the same result is an example of Einstein's definition of madness. They are like analogue broadcasting stations failing to adapt to the digital age, unable to win power through proportional representation under the post-devolution settlement, unsure of their relationships with their national parties in Westminster, demoralised and seemingly destined for perpetual opposition.

The clue is in the name, Alliance for Unity. We have to unite the split anti-nationalist vote, a wasted vote that has nevertheless been cast by the majority of Scots in every election since devolution, and that means an alliance. So far so good. But political movements don't get far without some unifying doctrine. As we found enthusiastic well-wishers in Scotland's academic community I formed discussion groups on Twitter that thought more deeply about what it was we were trying to defend, why it was increasingly being rejected by Scots, particularly the young, and we identified a need to redefine the unionist argument. The alliance bit is conceptually easy, albeit tactically very hard to achieve, though perhaps not as hard as everyone was quick to say a few months ago. It was the word unity that needed closer examination.

What is becoming clear is that, if we are to preserve the union, unionism alone will no longer cut it. We need to declare, 'Unionism is dead, long live Unity' and talk instead about the unity of the

United Kingdom, which is a very different philosophy, not mere semantics. Before I am excommunicated by my friends in the Conservative and *Unionist* Party and de-followed on Twitter by members of the Glasgow Rangers Supporters' Club, I should qualify that. It does not mean any less commitment to the union, it means more in fact. Nor does it mean any disrespect to the enthusiastic celebrants of the legacy of King William III, an important Scottish tradition that needs to be seen as cultural and local rather than political and national. But unionism as the prevailing orthodoxy has no future. There is no point in rallying round a culture that many feel excluded by. The separatists have cleverly used identity politics to spread sectarianism across Scotland and demonised 'Yoons,' so spreading a unionist message supports their narrative. More critically, the word 'union' implies two entities that have been joined together and the focus is on the linkage and how that must be tinkered with to make it work. Unity is a better word as it implies oneness, like two metals that have been melted down and then recast in one indissoluble alloy statue of Britannia.[51] This idea should be easily understood by Scottish families as it is also what has happened to our people. After three centuries Scots are fully integrated across the UK and throughout the Anglosphere. Blood is also indissoluble. Even Nicola Sturgeon has an English grandmother. Unity is therefore supportive of families and does not want to see them split. Nationalism, on the other hand, is toxic and is already dividing some families and would see many of them alienated from each other by the separation and partition implicit in 'independence'. At its most extreme, 'blood and soil' nationalism is a deeply racist doctrine that stokes Anglophobia.

This indissolubility implicit in the concept of unity should give us greater confidence to legislate to protect it. We should not be shy about saying that although the union was originally a marriage of two nations it is now the children of that marriage who should be our primary concern and they should be regarded as indivisible unless there are very exceptional circumstances. Our Clarity Act

---

51. Which is what in fact happened in 1707 when both parliaments voted to dissolve themselves and re-constitute as the parliament of one United Kingdom.

idea recognises the gravity of secession for many British families and frames the terms and conditions of any divorce, not so as deliberately to make it harder, though that is how separatists are portraying it, but to ensure that it is not undertaken wantonly or without due consideration. Appeasement of nationalists in the name of unionism has failed. Every time they have been given an inch they have taken a mile and they always come back for more. A unity approach would rediscover the power of No, much as the Spanish have done with Catalonia, and accept poor short-term optics and opinion polls by recognising the importance of our unity with robust legislation. What unity is not is federalism. There are many voices[52] calling for a federal solution but it would take us further down the slippery slope to separation. Federalism would not survive the first big defence and foreign-policy challenge. Covid has shown us that. And it certainly would not put the separatist genie back in the lamp any more than devolution has done.

But moves to strengthen the union, important though these are, reinforce the contractual view of it. There needs to be a compelling positive argument for maintaining the unity of the United Kingdom and Scots need to embrace a vision of Scotland in the union. Here again unionism falls short as it is invariably backward-looking. Recalling past glories and shared achievements resonates less with people now and hardly at all with the young. After thirteen years of ruthless control of the education system by the SNP the Scots are a people dislocated from their history between 1707 and 1945. There has been a carefully coordinated campaign to erase British culture from Scottish life and promote nativist Scottish culture. Remembrance Sunday was a case in point. Examples of the latter are everywhere you look in the bogus Gaelicisation of road signs. They have even quietly changed the colour of our national flag from Union-Jack blue to the lighter blue seen on Saltires flown at independence rallies.[53] Every sinew of government has been bent to their cause.

It has been highly effective but there are signs that they have overreached themselves and been so blatant about it that they have laid themselves open to ridicule and the charge that they have

---

52. Including ex-Prime Minister Gordon Brown's.

53. Although in fairness this is taking it back to its correct earlier heraldic colour.

allowed public services to deteriorate while they have pursued this narrow, divisive agenda. This agenda is a deliberate assault on British values that are still deep within most Scots, whether they realise it or not. Values that ironically derive from the Scottish Enlightenment such as fair play and tolerance and respect for truth and reason. The SNP have talked of freedom as if they are the true heirs of William Wallace while pursuing obnoxious authoritarian policies. It is as if they have tapped into the intolerant Knoxism that has always lurked deep in the Scottish character and seek to reverse the Enlightenment. They have consistently failed to win the rational argument for separation, unable even to produce a credible plan for the currency, so have instead appealed to base instincts and the false religion of nationalism. We are turning our backs on the age of reason and retreating into superstition. It seems as though it is only a matter of time before we start burning witches again. The evidence is also there, if it is looked for, to attack the SNP as being ruled by a cynical clique, using nationalism as a tool to seize power for power's sake and maintaining a perpetual 'neverendum', as the reality is that is better for them to travel towards separation than to arrive. It is all too easy to delineate the careful construction of a one-party state from the opportunities occasioned by the introduction of proportional representation and a unicameral parliament.

A change from unionism to unity would press the reset button and allow us to use a different narrative. It should explicitly focus on the family and family values of truth, decency, tolerance and honesty, which are also part of the British character, a fusion of English tolerance and generosity with Scottish morality and intellectual rigour. There is already widespread revulsion at the more extreme policies of the SNP and at the corruption that is becoming more apparent every day. We should reject nationalism as an aberration. We should call out post-truth fakery and identity politics as part of a Counter-Enlightenment. And Scottish writers and thinkers should take the lead in promoting a Re-Enlightenment in the way that Hume and Smith once did. And we should spread a positive message of individual liberty and neighbourliness.

After thirteen years of unrelenting nationalism Scotland should be receptive to change. The paradigm shift caused by Brexit and Covid forces a re-evaluation. There is good reason to be deeply suspicious of the Great Reset agenda of the globalist elite. But we will need a recovery, and therefore some fresh thinking and re-orientation. We also need to be honest and acknowledge that this aberrant nationalism, like Trumpism and aspects of Brexitism, has happened because there is a large section of Scottish society that feels left behind by globalisation and abandoned since de-industrialisation a generation ago. Scexiteers and Little Scotlanders are the tartan-clad equivalent of Brexiteers and Little Englanders. They need to feel included and that life will get better.

This recovery has to be approached from a position of unity, not because Scotland is 'too poor or too wee' but because the challenges are so massive that we will need to harness the collective human capital of the whole British people as well as the financial capital of the Bank of England in order to thrive. We are all in it together, and we need to hang together in a globalised world or be hanged separately. One of George's first acts was to choose the Spitfire roundel as the symbol of the Alliance for Unity, a deliberate reminder of the Battle of Britain, a very British symbol but one without the jingoism of the Union Jack, or the 'butcher's apron' as separatists call it. It is a reminder of what Britons can achieve together against all the odds.

We should emphasise that Scots have led the way in Britain before and we need to be in the forefront of this recovery. Indeed Air Marshal Dowding, without whom the Battle of Britain would have been lost, was a Scot. It should not be a hard sell. Unfortunately the political capital that should have accrued to the unity argument from Rishi Sunak's millions and the Oxford vaccine project has been squandered in the short term by yielding the floor to the First Minister at a nakedly political daily press briefing.

Above all, unity transcends party politics. It is neither left wing nor right wing. It draws on the socialist idea of solidarity across the United Kingdom, on similar one-nation Tory ideas like support for the family, and on Liberal anti-authoritarianism. The Alliance for Unity is a coalition of people from across the political spectrum. If our candidates can reconcile our very diverse political ideologies in the face of the existential threat posed by separatism so too could our parent parties. But will they?

* * *

Last week it was my privilege to stand in glorious parkland, studded with ancient oaks, chestnuts and beeches, and shoot pheasants. The leaves were at their exuberant best with the palette of reds, yellows and oranges at its most fiery in the gloom of a dreich November afternoon. But the stars of the show were the larches flecking the giant stands of conifers with gold on the hills above us. Our host, a keen and knowledgeable forester, as many Dumfriesshire lairds are, bemoaned the fact that they had succumbed to the deadly *Phytophthora ramorum* and were due to be clear felled. Each year there are fewer of them as the deadly disease takes its toll. The landscape will be changed utterly when the backdrop to our world is a monochrome dark-conifer green. He is not going to replant with any larches, nor would he be granted permission to do so. There are doom-laden articles blaming the import of nursery stock 'from the Continent'. They are right in one sense, the fungal disease probably has come in from abroad; it had to come from somewhere. And it is a lesson for our political masters about the dangers of importing anything animal or vegetable without a rigorous process of checks. But the larch is not indigenous, not in this inter-glacial period anyway. The European larch was first imported 'from the Continent', specifically from the Tyrol in the Austro-Italian Alps by the Duke of Atholl to plant on the hillsides around Blair Atholl in Perthshire in 1738. And its Japanese cousin sometime later. Now that the demise is endemic there is not much we can do about it except fell any diseased trees. The demise of the larch will not be good news for sporting estates. The larch, the only deciduous conifer, is responsible for letting the light into many a good coniferous pheasant covert in winter. And a screen of larches around the sides of woods aids the flighting of pigeons at dusk. It is also the best firewood for my biomass boiler with a higher calorific value than ash. We will miss it when it has gone. In the meantime we should appreciate it.

* * *

We have finally sold all our farm machinery. Covid delayed the sale by two months. The Council vacillated until the weather broke before finally giving permission for it to go ahead, so that we trudged around in mud. Everyone loves a farm sale. The auctioneers had predicted that farmers wouldn't be able to resist coming to have a snoop at the new dairy and so it proved. I was asked more questions about what was in our grass leys than I was about the machinery. And the new buildings were assessed by several hundred pairs of critical eyes. Farm machinery sales, like house sales, are often preceded by death, divorce or debt and they are for us all a premonition of what might happen 'if it all goes tits-up'. Happily in our case it was just occasioned by a change of direction, but I had several worried enquiries from friends who saw the advertisement and assumed the worst. The sale followed a predictable routine with several chancers coming round in the preceding days offering me good offers for cash 'to save having to put it in the sale and risking getting less for it'. They were all told to turn up and bid and happily all the lots exceeded their 'best prices'.

CHAPTER 12

Party Politics

Commentators on Scottish politics have their tribal allegiances like the rest of us. That is why they tend to be very dismissive of any new parties threatening to muscle in on the comfortable three-party opposition at Holyrood – and opposition is what it is and likely to remain without a radical change of attitude. They prefer to cling to the dream that their leader is going to lead the long-awaited recovery in the Conservative or Labour vote and sweep the nationalists from power. The Liberals need the opposite to happen, a failure of either party to command the electorate's trust decisively so that the vote is split in such a way that they can come through the middle and be the kingmaker in coalition. All three parties and their cheerleaders are stuck in the pre-devolution politics. They are ignoring the SNP elephant in the room. The polls have been consistently saying that the separatists have a solid 45 to 55 per cent of the vote. And so the old parties are splitting around 50 per cent of the vote three ways by failing to embrace coalition BEFORE the election rather than afterwards.

At the other end of the spectrum are the smaller parties. They are all too aware of the changes to the electoral system and view the proportional representation component of it as an opportunity to further their (usually single issue) ambitions. The toxic nature of Scottish politics right now, and the disenchantment with the failing opposition parties who have allowed extreme nationalism to rise before their eyes, leads them to believe, with some justification, that there is a constituency of angry people out there who will give them their protest vote. Thus there are at least two parties with a commitment to abolish the Holyrood parliament. And there are parties that are anti-lockdown or pro- or anti-Brexit and therefore hoping that matters will still be unresolved with Europe or Covid. Now there are rumours that Nigel Farage and Richard Tice are hoping to enter the fray, hoping that Scots are ready to embrace their own distinctive brand of right-wing English nationalism. I ring Tice to try to persuade him to stay out of Scotland or, if he really is determined, to come into our alliance. He is not prepared to listen. I also ring Laurence Fox, worried that his Reclaim Party might also dilute the anti-nationalist vote. Laurence gets it straight away and is supportive of what George and I are trying to do. He agrees not to stand against us.

There has been an attempt by some commentators to lump Alliance for Unity in with the small party end of the spectrum. The narrative fed to tame journalists by party spin doctors seems to be that we are 'unlikely to win many votes' and 'only likely to split the vote further'. But we are already an alliance of several of the smaller parties who have taken the pragmatic view that in order to have any chance of promoting their own views in a coalition government it is vital first to defeat the SNP and install a government of national unity. Thus we have united the vote rather than split it as the Scottish Unionist Party, the Workers Party and the Scottish Christian Party have already parked their planes on Alliance for Unity's inclusive aircraft carrier.

Unfortunately – and utterly bizarrely – the Election Commission doesn't see it that way. This shadowy, unelected quango, which wields huge powers, for example on what parties can be registered to contest elections and what questions can be put in national referendums – won't register us unless we change our name. They don't accept that we are an 'alliance' despite us being probably the most diverse alliance of different parties in British political history. We propose a compromise 'All for Unity' but they turn that down as there is apparently a Unity for All Party. This is farcical as it turns out to be a one-man party run from a bedsit in Croydon and has nothing to do with Scotland. Eventually, after prolonged negotiations by our brilliant young organiser James Giles, they agree to register us as All for Unity. But the delay has hurt us badly. It has meant the media has ignored us as we have not been registered and it has damaged our reputation and deterred donors each time the Electoral Commission has very publicly turned us down.

It is very depressing that the one organisation that has the power to facilitate pluralism of political parties, an important democratic safeguard, has been our biggest obstacle.

* * *

The government's preliminary announcement on future farm subsidies fills me with gloom. I reluctantly voted Remain with my head because I felt, cynically and selfishly, that British politicians would never protect farmers' interests as well as Irish and Continental ones. But in my heart I hoped that the Brexiteers would accomplish the radical New Zealand style transformation of British agriculture needed to free farmers from the tyranny of state control and allow them the opportunity to earn a fair wage from the global market place. George Eustice's template for English farm support – we don't yet know what is on offer in Scotland[54] – looks depressingly like EU-lite. Payments to farmers to maintain the environment are to be replaced with, er, (smaller) payments to farmers to maintain the environment, nothing more than a slight change of emphasis. For the last twenty years I have been receiving payments for hedges, ponds, rushy pasture, water margins, wild-flower meadows and winter stubbles. I committed to various 'schemes' because I view wildlife conservation on my farm as just as important as the food we produce. The payments have been miserly, never quite enough to compensate for the loss of the income I would have received if I had farmed more intensively and certainly not enough to compensate for the form filling and the stress of long-running battles with civil servants over interpreting ever-changing rules. The final straw came when I was made to keep a diary like a primary school child. I have come to the conclusion that it is better to farm for maximum profit and use any surplus for conservation on my land than to be a poorly paid serf of the Green State.

---

54. And disgracefully we still don't at the time of editing in April 2022.

The underlying thesis of post-Brexit Britain is now the same as the EU's: the patronising notion that farmers should be paid just enough to stay in business in order to curate the countryside within carefully prescribed and frequently illogical guidelines, while food leaves the farm gate at or below the cost of production, helping to boost the profits of quasi-monopolistic food processors and supermarkets.

Farming charities cite cash-flow difficulties and red tape as the primary triggers of mental health breakdowns in farmers. This is not going to be helped by less cash for more tape, even if the tape is green not red. Those of us who have farmed through previous subsidy cycles know what is coming next. Cue the hoop industry. Farmers will need to jump through certain hoops in order to qualify for support. A hoops roadshow will take to the highways to consult 'stakeholders' about the design of the hoop and, more importantly, the hoop compliance regime that will require legions of civil servants to inspect the hoops and fine the farmers for not jumping high enough. Process will take priority over outcomes. The hapless farmer will find that as just one of many stakeholders he will be accorded parity at best with the noisy demands of the green quangos and charities. The taxpayer will be ripped off because the administrative cost will dwarf the sums that eventually filter down to farmers in subsidies. The 'progressive' social engineering implicit in the 'small farms good, big farms bad' approach to the tapering of future subsidies sounds good politically but will in reality just mean that medium-sized farms like mine will have to farm more for profit and less for nature.

The Brexit vision, articulated during the referendum campaign by Owen Paterson and others, appears to have been hijacked by the Green Blob. The radical thinking by Tim Lang, about restructuring our food system from plate back to field, quietly shelved by the civil servants. Talk of de-regulation and fair competition seems just to have been talk. Is this really what Brexit was all about?

December 2020

December birdwatching is a bit hit and miss. Nothing stirs then all of a sudden the air is filled with birds like that Hitchcock movie. The songbirds have flocked together to protect themselves from the hawks and move around the farm in a great chattering phalanx. The starlings split into platoon-sized patrols to feed on our fields by day before assembling each evening for a brigade-sized murmuration at their roost two farms away. And the geese either fill the sky in their thousands or are largely absent. So we were very pleased with ourselves when we took some friends for a walk along our foreshore after a socially distanced lunch at the pub on Sunday and went through the bird bingo card. The tide was high but looking-glass smooth under a mother-of-pearl pink sky that turned steadily orange into the gloaming. And the waders flew up and down entertaining us with their musical acrobatics. Curlews and oyster catchers piped us down the beach and a small flock of knots mesmerised us with their silvering aerial kaleidoscope. The geese briefly stole the show as they flew out to roost. But the best sight of all was the lapwings. The collective noun for lapwings is either a desert or a deceit. Desert seems an odd word for them although sadly now appropriate, as 'peewits' are becoming increasingly rare. Deceit is spot on though. Anyone who has seen them putting on a display to trick predators around their nests will understand this. But in the eerie half-light their deceptively ungainly flight caused them to flicker white then black (or actually dark green) as we saw first their bellies then their backs rendering them partly invisible, a brilliant deceit.

* * *

As the nights draw in we are missing our friends even more. By now the winter round of dinner parties and bridge over kitchen suppers would be in full swing. I have been reading a recent paperback reprint of *A Distant Drum* by Jocelyn Pereira. It is one of the best memoirs of the Second World War by an officer in my regiment, the Coldstream Guards. The regiment can claim to have produced the three finest wordsmiths of the war: Pereira, Professor Sir Michael Howard (author of *Captain Professor* inter alia) and Roger Mortimer, later to win posthumous fame as the long-suffering Dad in *Dear Lupin*. Mortimer's letters from a POW camp are in *Vintage Roger*. The three of them should be read by anyone whose spine needs stiffening during the current emergency or those who make the mistake of comparing it to a war. I have been bolstered by curling up at night knowing that I am not lying under a greatcoat at the bottom of a trench in Normandy during an artillery 'stonk'.

* * *

After the most socially boring year ever we can't wait for the family gathering at Christmas. It will start in time-honoured tradition with the hunt for a tree and the greenery with which to festoon every picture and mirror in the house. This involves piling everyone into the pickup and trailer and heading into the woods with chainsaw and loppers before fanning out to look for the perfect tree, holly berries, ivy and, if we are very lucky, mistletoe. It's traditionally the setting for the First Family Argument of Christmas, I would write a parody carol if I could. Suddenly every member of my family becomes a world expert on forestry. Some years it has been nearly dark before we settle on the best-shaped tree to cut down. The Executive Producer of Christmas is usually the most vociferous and sometimes the divorce courts are not far away. Fortunately a few years ago I hit on the idea of settling the argument, and in doing so attempting to instil the rudiments of silviculture in the younger members of the clan, by making it a rule that the tree has to be one that has to be culled for the greater good of our woodlands because it is misshapen or growing too close to another one. Recently, because of my aversion to planting spruces, we have struggled to find a young one the height of our drawing room ceiling so we have resorted to felling a larger tree and cutting the bottom off. This has introduced a conservationist counter-culture to our environmentally aware Christmas and 'Dad's wonky tree' is a talking point and sometimes achieves a cult following among the children's social media followers. They all look the same once they are covered in tinsel anyway.

* * *

The green manures have been covered in seeds all winter and consequently pigeons, finches and ... pheasants. This hasn't always been a nice problem to have as far as the shoot is concerned. Trying to deliver six evenly balanced drives isn't easy when the birds all want to be in one place. And some of the fields don't lend themselves to being sporting pheasant drives. But it has introduced a bit of variety into the shoot with a new challenge each year and shown us places that might be good locations for game crops in future. This year's layout has worked the best so far. The beaters put up clouds of songbirds, mostly goldfinches, coming off and making me wish that I could shoot with my binos hanging round my neck (they always get in the way whenever I try). Then a flock of feral pigeons that roost in a nearby steading have flirted with the guns with reliable predictability by doing a couple of circuits. Then there has been a steady stream of pheasants with some high birds testing the guns. Next season will be back to the drawing board though.

The pandemic has made shooting harder but the spirit of 'keep calm and carry on' has prevailed, evoking memories of my grandmother who took up shooting during the war along with many of her friends. Photographs survive of them doughtily shooting in tweed skirts to feed the neighbourhood while old men and boys, and no doubt Italian prisoners of war, did the beating. I am enjoying the novelty of shoot picnics on car bonnets and in open-sided hay sheds. Best innovation has been the pheasant pasty patented by our new tenants Jamie and Sarah Brighty.

January 2021

The Chinese must be laughing their socks off. Having unleashed a deadly virus on the world, knowingly or not, they now have the spectacle of ten useful idiots in the Tory Party setting up 'The Vegan Conservatives'. Their 'spokesperson' Andrew Boff is 'thrilled that so many Conservative MPs are going vegan for January'. He claims, 'At the heart of Conservatism is a desire to conserve and protect our environment. Moving towards a plant-based food system is critical if we are to prevent dangerous climate change, *reduce pandemic risk*, and protect animals.' Boff goes on, 'The modern vegan movement was born here in the UK, and as Vegan Conservatives we want to build on, and expand, this proud British tradition.'

Orthodox, omnivorous conservatives – once we have recovered our equilibrium – should be robust in condemnation. And we should use the publicity caused by this outbreak of food-wokeism in the Tory Party to have a serious debate about diet. Because Covid-19 has exposed a glaring deficiency in our immune systems. We have some of the worst Covid death statistics in the world. We have had to endure heartbreaking losses, sacrifice our economy and restrict our liberties because there are too many middle-aged adults suffering from obesity and type-2 diabetes. It is the pandemic that dare not speak its name. We now spend more on treating obesity related problems than we do on the police. And that is a result of eating too many carbohydrates and polyunsaturated vegetable oils and not enough saturated animal fats and green vegetables. Big Food makes us ill and Big Pharma cures us. Veganism exacerbates this cycle.

China's response to the pandemic, incidentally, was to tell their people to consume 300g of dairy products each day to boost immune resistance. An entirely logical response that would once have been part of the Tory armoury before they became obsessed by neo-liberal gesture politics and afflicted by a strange myopia when it comes to metabolic health.

* * *

The frost has been, on balance, a good thing. The ditches are still running well after a week or more without rain, which is a sign that the daily freeze-thaw routine is cracking the ground and letting water down into the field drains. The earth is squeezing moisture out like a dirty sponge and our soil structure will be that bit more absorbent come the next wet spell. We might even be able to get some tractors on the land to tidy up the hedges. It has also helped the team at the hunt kennels to be philosophical about the lockdown. They couldn't hunt anyway with the ground this hard. The shoot, on the other hand, is poleaxed by the First Minister locking us down and my morale is not helped by hearing record numbers of cock pheasants cock-cocking up to roost in the main drives, drawn in to the feeds by the hard weather from the fields and hedgerows where they had been wandering. We will probably have to concede defeat to them now and treat them as pets, as we feed them through until harvest time, in the same way as we do the red squirrels and songbirds that come to the bird table.

It was all going so well. We managed four of the six days I had planned. We had embraced the whole, alien Covid modus operandi of masks in vehicles and outdoor lunches, and not stroking dogs or handling game, and tips in envelopes. And against the odds we snatched fresh air and fellowship and high, curling pheasants that will now linger in the mind longer than they would otherwise have done. Then the order, counter and disorder from London and Edinburgh, the emails and telephone calls and the growing realisation that the season is effectively over.

* * *

The hopes and fears of all the years are focused on the heifers in their smart new sheds. It is three and a half weeks until C-Day when the first calves arrive and the first tanker load of milk leaves for the cheese factory at Lockerbie, and the excitement is growing. The parlour gleams expectantly, bristling with steel pipes over freshly laid concrete. There are still several installations to complete before we can pull the first pint and we are praying hard that none of our 'key workers' suddenly have to self-isolate before they can finish it.

One silver lining of the pandemic is that the vegans seem a little less vociferous this year, when there is a much bigger health story dominating the headlines. I have wound them up on Twitter a couple of times but the fight seems to have gone out of them. Perhaps a lack of meat in their diets has taken all their energy away. The fear at the back of every dairy farmer's mind is that we will wake up one morning to find the world is drinking almond milk lattes. It takes a leap of faith to remember that for every virtue-signalling pre-osteoporosis case shopping in the politically correct foods aisle, there are probably two across the developing world who want the nutritional value of the real thing. That's what I keep telling myself anyway.

* * *

The year 2021 started with a barrage of separatist propaganda issued through official government channels. A taxpayer-funded video called Brexit 'a bad deal for Scotland' and asserted 'Scotland has a right to choose a better future as an independent country.' Public money continues to fund Covid advertising that carries the subliminal message #WeAreScotland. And the BBC continues to allow a party political broadcast in all but name every afternoon within months of an election. The Saltire, which belongs to every man, woman and child in Scotland, has been allowed to become a symbol of separatism like the Irish Tricolour. This conflation of party and state would never have been allowed even ten years ago. The independence of our civil service and the BBC were once admired across the world. It is now widely assumed that our civil service and large sections of our broadcasting media have been politicised and this corrodes trust in the machinery of government.

Everywhere you look there are signs of corruption, waste, negligence and incompetence – rusting ferries, unbuilt hospitals, falling educational standards. It takes two days for the Isle of Man to jail a jet skier but Margaret Ferrier still has not been in court.[55] In any other democracy a ruling party with so much failure and dirty linen after fourteen years in power would be facing defeat and an eager opposition would have all the appearances of a government in waiting. But that's not happening either. The three opposition parties do enough to justify drawing their MSP's salaries but not enough to dent the SNP's poll lead or to prevent them behaving as if we live in a one-party state.

***

I wake to a sea of grey, literally. A party of oyster catchers is huddled on the tideline waiting for the ebb to uncover their breakfast, blending perfectly with the seaweed in the dawn half-light. The frost has gone and with it the sunshine and the snow-capped fells of the Lake District across the Solway. Yet again I have missed

---

55. The jet skier had broken the Isle of Man's Covid lockdown by skiing across from Scotland. Margaret Ferrier MP had admitted to knowingly taking Covid into the Palace of Westminster months before.

the chance to get tractors into the fields to trim the hedges while the ground is hard. We always try to do this after the birds have eaten all the fruit, but frosts are increasingly rare and we have now missed an opportunity. When I was hunting regularly the end of a frost was cause for celebration but the hunt is grounded by Covid and driving past the kennels I can sense the hounds' frustration and hear the occasional yelp as they take it out on each other. No, there is little to recommend the return of milder weather, except, perhaps, that I won't be having to stoke the boiler quite so often to keep the Thermostat-in-Chief happy.

The departed frost will leave us with wonderful family memories of treasured and all too rare time spent with Oliver and Rosie though. Over the holiday we walked our way through Dumfries and Galloway's best walks. Each one had special serendipities. Along the cliffs near the Isle of Whithorn we saw a big dog otter slither out of the foam in a tiny inlet and search for crabs on the rocks below, and held onto the dogs tightly as they threatened to pull us over in their excitement. At Balcary Point we watched a school of porpoises lazily breaking the surface with their fins. And we braved the ice and clambered up a slippery path to the top of Screel Hill and saw the sun set on the Solway coast with half of Galloway laid out below us, perhaps the finest view in Scotland.

Walks nearer to home have been accompanied by the cat. Yes, the cat. Those who know me will know that I grew up in a household that held firmly to the dictum that the only good cat is a dead cat. This had unfortunate diplomatic consequences on occasions when our gamekeeper shot cats that subsequently turned out to belong to our tenants. Some of the ensuing feuds went on for decades. I have modified my views somewhat having spent many thousands of pounds rewiring properties after mice have chewed through the cables. It has rarely happened when there has been a cat in the house. So I have become pro-cat in winter but anti-cat around Easter when the mice go out into the fields again and the birds become vulnerable to the feline menace. If I had my way we would round up all British cats in the spring and send them to New Zealand and bring them back in October. But this thawing in my cat attitude had not developed as far as to get one. I blame the children – who are still disinherited if they are reading this – for rescuing 'Ibble' as a kitten from her family home in the bottom of an old ash tree we were felling, along with her brother 'Dibble', who has mercifully now been found a loving home elsewhere. Come to think of it, the house hasn't smelt of dead mice on the stairs so much this winter. And Ibble does have a certain charm and, being black, may bring luck every time she crosses my path ...

Luck is what we need now that everything is prepared for the new dairy venture. We have had our first calf, born prematurely, a healthy heifer, which is what one wants in a dairy herd, and the cows are all 'bagging up' now and it won't be long before our young team swings into action calving 600 of them. It is a cosmopolitan team, all under thirty: Joey our manager, Dan the second-in-command, Scott, Olly and Yuri are respectively an Irishman, a Welshman, a Scot, a Devonian, and a Dutchman – not what anyone expected after Brexit.

One person who is an avowed cat lover is George Galloway. George grew up in Dundee and has never lived in the countryside before. At our meeting this week he is wide-eyed with excitement at having just witnessed a kestrel stooping on a vole. We will make a countryman of him yet. Czechoslovakia had a velvet revolution, Scotland's tweed revolution edges closer each day.

February 2021
The days are lengthening – though, as we have not seen the sun for a week, it doesn't seem to make much difference under murky winter cloud. Skeins of geese are buffeted across the skies, strangely silent unless you happen to be downwind of them and even the dogs are uncharacteristically reluctant to venture outdoors. Still, there is hope. I have a simple winter algorithm in my head that keeps my spirits up: I count the number of days back to the solstice then double it. That way you wake on Christmas morning knowing you have already had the darkest week of the year. The snowdrops are doing their best but seem a little behind most years and have yet to turn the woodland floors white and, though it might be wishful thinking, there do seem to have been more birds going around in pairs. By contrast the roe deer have been coming together. Twice I have seen groups of six or seven grazing together. The shelduck have appeared on the shore again, which is usually one of the first

signs of spring here. But otherwise there are few signs that the natural world shares my optimism as yet.

This month's wildlife excitement has been frequent sightings of otters close to the shore. They appear to be taking up residence in the burn that forms the end of our garden. I have a sneaking suspicion that otters are nearly back to pre-crash population densities in this part of the world. A fact that is probably directly related to the decline in mallard numbers. Dumfries and Galloway has a long and honourable tradition of otter conservation. The Dumfriesshire Otter Hounds was one of the first organisations to notice there was a problem and push for their protection. Getting governments to agree to the control of this lethal predator one day might be a different matter.

An otter was first seen by our friend Sophie, who is a 'wild swimmer', swimming in the sea at high tide in all temperatures. Each to their own. Fortunately an otter appeared on the beach in front of the house while she still had her clothes on and she was able to film it foraging along the tideline and post it to YouTube. The footage is now being shared on social media as part of our advertising drive to get the holiday cottages filled if, or hopefully when, the lockdown restrictions ease at last. Predators have their uses.

CHAPTER 13

Milking

There is a moment in every entrepreneurial venture when the agonies of planning a new business, the frustrations of working in a muddy building site and the yearning for it to be real are suddenly replaced with the fierce urgency of now; and there is barely time to savour it because operational life is even busier than development life and, far from being an arrival in sunlit uplands and life becoming much easier, there are suddenly a hundred and one things to be attended to. Or in our case living things – Jersey–Friesian cross heifers needing to be calved then milked. But savour it we did. Corks popped, a ribbon across the entrance of the milking parlour was cut by Sheri and the team drank to our mutual prosperity in the biting cold of a January evening.

The whole project had been planned for a calving date of 1 February. That way the whole herd should be reaching peak lactation by the time the spring grass comes. The only problem was that one of the farms in Ireland where we bought some heifers had clearly let the bulls out earlier than they thought. The first calves started appearing in abundance a fortnight early before the milking parlour or the calf shed were quite ready. Improvisation has seen us through with the concreting team just managing to stay one step ahead of the spreading carpet of straw dotted with calves curled up asleep like oversized cats. While hasty last-minute connections were made by plumbers and electricians to allow the parlour to start extracting precious colostrum, followed days later by our first tanker load of rich, creamy milk. I love it when a plan comes together.

When we first sat down to design the new milk-from-grass system I was ready to embrace the latest ideas about paddock grazing but there was a lingering doubt in my mind about whether we really could have cows out grazing in February. But there they were in the last week of January, happily munching away in one of the drier paddocks before walking back along the cow tracks and into the parlour. It has been all the more satisfying knowing that most other dairy farms in the country have been covered in snow but no doubt this guilty pride will come before a fall and we will have our setbacks.

Cows being milked form some of my earliest memories. The sensual synthesis of steaming cows and frothy milk bubbling through the pipes, the dull rhythm of the pumps, and soft words of encouragement from the dairymen punctuated with the splatter of 'skitters' and muffled curses, and musky cow odours mixed with acrid smells of muck and urine overlaid with the sweet aromas of milk and cattle cake. The whoosh of the cake through the pipes to the troughs is still the same, conveyed from the feed tower by the miracle of augurs, which are still as much of a mystery to me now as they were when I was a child. It seems as though nothing has changed and the forty-five years with no dairying here were but the blink of an eyelid, which, in the long view of farming, I suppose they have been. It was a great childhood treat to be allowed to go and watch milking and it is a happy thought that one day I might be able to take grandchildren to see it.

The boys in the dairy team are amazed when I tell them that I can remember cows being milked in byres. The latest innovation then was the bulk tank, which replaced the scores of churns left at farm road ends, and I can remember being lifted up to look inside and watching the paddle going around, and ladling fresh milk into a small churn for the house. Now you would need a ladder and the tank, which is the size of a lorry, is sealed anyhow for very good hygienic reasons. But there is a tap where we can draw off our milk. We still have the churn but my father kept creosote in it after we gave up dairying, so I don't think we will use it again.

The other treat was being taken to feed calves. Again, the memory of small, dark calf pens – now converted into hound lodges – contrasts with our new calf shed, which is as clean and bright as any paediatric ward. I am becoming accustomed to being castigated by more sentimental friends for the implied cruelty of removing dairy calves from their mothers. And, in truth, I do regret the loss of our suckler herd with its overt displays of maternalism. But a heifer is very often clueless about this strange creature that has appeared painfully from inside it and in fact it is quite common for newborn calves to be trampled. The bonding takes time and if the calf is whisked away to be given colostrum and put to bed in clean straw the loss is not so great and actually far less than the agonies at weaning time when the beef cows used to bawl for days like mothers in a prep school car park.

Amid the excitement of calving and the appearance of new life it's tempting to think it is spring. But one look across a polar sea at the snowscape on the Cumbrian Fells dispels the notion. Flocks of twites are swirling around on the seed mix fields, which must seem like a well-stocked larder to the birds amid the bare expanse of grass everywhere else in the parish. They are occasionally chased by the merlin. Merlins and twites are natives of the heather moorland and their annual move to the seaside to continue their warfare is the Scottish equivalent of lions following the wildebeest migration. There is a renewed urgency to the birds' feeding on the bird table and the geese have reappeared on the Solway in force, driven south by the snow. The hungry season is starting to bite. It is snowdrop time though and the woods are bright with their drifts, a tribute to the dedication of my grandmother who divided them devotedly each year and spread them far and wide. They are our annual sign of hope and better things to come.

***

I used to be fond of Scottish banknotes. I enjoyed the banter with London cabbies who refused to take them. As a child growing up in Scotland I was proud of their Scottishness that differentiated us from our bigger neighbour. I liked their quirkiness and the very British anomaly that the Bank of Scotland, Royal Bank of Scotland and Clydesdale Bank have, for arcane historical reasons, a licence to print money (the origin of the expression).

But now I would dearly love to abolish them. There comes a point when an anachronism ceases to be amusing, when it really matters. And it does matter now because, just as the future of Great Britain as we have known it for the last 300 years starts to hang in the balance again, Scottish banknotes foster the illusion that Scexit would be easy. Nationalists speak breezily of Scotland's prosperity once we are free of the shackles of Westminster. In unguarded moments, when probed on the weakness of their case, they say that, if we are short of cash, 'We will just print it.' They talk about Scottish pounds as if they are 'a thing'. They feel entitled to keep the pound post separation because, 'It's oor poond.' These are dangerous fallacies.

Besides, everyone knows that when they read on the front of an RBS banknote that its Chief Executive promises to pay the bearer, on demand, the sum denoted, it doesn't mean anything of the sort as RBS is bust, dead, deceased, passed on, no more, ceased to be, expired and gone to see its maker, bereft of life, an ex-bank. And the fact that the banknote is worth anything at all, and is not just Monopoly money, is because the British taxpayer stood behind RBS when it went spectacularly tits-up in 2008 along with its counterpart the Bank of Scotland, which now exists merely as a brand within Lloyds. It is no more a bank now than the British Linen Bank, which the Bank of Scotland itself acquired in 1969. The Clydesdale Bank had ceased to exist as an independent entity long before that and is now part of Virgin Money.

Scotland does not have a genuine banking sector any more. It went west in the banking crash (under the supervision of a certain Scottish Chancellor, but let's not go there). The relationship manager at my bank told me that the three Scottish banks would never have their brands absorbed by their parent banks, as the licence to print money is seen as great advertising. So they continue to pose as real banks with real reserves on Scottish high streets and the illusion is linked to the licence.

By supreme irony the chief proponent of the SNP's economic policy post-Scexit, if it ever happened, is none other than Andrew Wilson, who was formerly Fred 'the Shred' Goodwin's right-hand man at the Royal Bank of Scotland. We all saw how the hubris of the bank's financial projections worked out when the proverbial hit the fan in 2008. And it didn't end well. Wilson's Charlotte Street Partners pump out rosy predictions about how we Scots, currently part of a G7 nation, might aspire to be a prosperous country like, er, Denmark … after, ahem, a generation. 'Charlatan Street' came up with the wheeze of using 'sterlingisation' to transition to a new Scottish currency as a compromise to hide the gaping lacuna in the SNP's plans for a currency post-independence – a policy that has since been rubbished by most respectable economists.

Now that I find myself as a dairy farmer with a large bank loan in sterling, I take a dim view of the nationalists who would make me one day attempt to keep up the loan payments in a different, probably devalued currency. And as an army pensioner who is ten years away from a state pension, I do not want to spend my declining years worrying about where the money to pay my pensions is to come from while the separatists fumble their way to those Danish uplands.

Our Clarity Act would provide greater transparency and remove any grey areas, making it less likely that my countrymen would sleepwalk into a disastrous Scexit. Clarification of banknotes would be a good first step. If the Prime Minister is serious about saving the union he should embrace the doctrine of unity rather than unionism. He should rename the Bank of England: the Bank of Britain – it was founded by a Scot after all – and insist that there should only be British banknotes. They could, as Bank of England banknotes already do, have Adam Smith, Scotland's greatest economist, on the back.

* * *

The tide is in under a cold, clear sky and consequently we are blanketed in a thick haar, or sea fog: this is the natural equivalent of running a hot bath in a cold bathroom. They are more of a feature of life on the east coast but we get them in the mornings on the Solway sometimes as well. It is frustrating as I know that a mile inland it will be a lovely day. The sun's attempts at burning through are like spring's nascent efforts. It seems tantalisingly close – I broke out into 'shirtsleeve order' for the first time a few days ago – and there are signs of life stirring everywhere. The

pigeons are paired up, gliding around contentedly like smitten
lovers and cooing in the woods. The fields are greening up quickly
after the milking herd has been on them. The cows are going
out to graze immediately after milking, but coming in after they
have eaten their fill to avoid damaging the pastures. The spirit of
optimism seems greater this year than most. All being well, the
spring blossoms should coincide with the easing of lockdown.
Holiday makers certainly seem to think so. Our holiday cottages
have never been so booked up and our inboxes buzz with cheery
emails looking forward to better times.

Zoom calls have taken off during lockdown and I am bullish
about the positive impact they are already having on rural life. More
people now see that they don't have to live in a city to work in 'the
knowledge economy'. The impact on house prices for those who live
off the land will not be so benign and I hope the politicians have
hoisted in the need to build more to cope with this extra demand.
One definite positive has been the ability for countrymen to interact
directly with politicians without leaving their farms. I have been on
several Zoom calls to lobby politicians on agricultural policy. And it
has been a joy to see a screen full of farmers in their everyday clothes
berating a squirming politician in a smart suit, in a range of rural
dialects, straight from their kitchens.

* * *

Times and *Telegraph* readers may have heard of All for Unity
because the newspapers believe in plurality of political parties as
a fundamental principle of democracy and have given us a fair
crack of the whip by publishing several articles from me. But if
voters only watch the BBC they probably will not have done as the
Corporation seems to have forsaken any pretence of balance.[56] It is
just one disturbing feature of the dystopian country we now live
in, where the conventions of public service, and especially public
service broadcasting, have been suspended for the perceived greater
good of 'independence' in the minds of the Scottish establishment.

---

56. It was a frustration that GB News, where George and I have been regular
   interviewees since the election, was not in operation until it was too late for us.

At Sandhurst the checks and balances of our constitution were drummed into me: the separation of powers between the executive, the legislature and the judiciary, the rule of law, the independence of the police from the state, the freedom of speech and of the press, the apolitical impartiality of public servants. I never dreamt that I would ever find myself, like many other All for Unity candidates, one day moved to enter politics to fight for them. Every principle is either threatened or has already given way to the SNP's ruthless building of a client state over the last fourteen years.

The Salmond Inquiry brings it all into sharp focus. We see that Nicola Sturgeon and her husband have been accused of perverting the course of justice to frame Alex Salmond. My impression is that parliamentarians, the judiciary, the media and the civil service have somehow managed to reduce any investigation to a farcical inquiry into procedural niceties chaired by a friend of Nicola Sturgeon's, Linda Fabiani, who had been sacked by Alex Salmond.

With hindsight it is easy to see how the complacent Blair government set up Holyrood without any of the checks and balances that would have prevented the corruption of power. For example, parliamentary privilege was not extended to Holyrood so questions may not be asked without exposing the inquisitor to a libel action. And how a political class of brash young people in suits, seemingly differentiated only by the colour of their rosettes, for whom politics is a career rather than a vocation, has overseen the decline. It is little wonder that the biggest responses All for Unity receive on social media are to our promises to 'open the books' and 'clean out the stables'.

Change is vital to restore standards of public service in Scotland.

***

'Morning Tory scum. How is our class traitor today?' I ask. The gravelly voice at the other end of line chuckles companionably. George Galloway is unfazed by the ritual online abuse he has been receiving from 'Natz and Trotz', as he described them in a tweet, after his announcement on Twitter that he will be lending his first vote to the Tories in the May elections for the Holyrood

Parliament. For a rational man like George, who sees nationalism as the antithesis of socialism and 'Scexit' from the UK as a clear and present danger to the poorest in Scottish society, who would suffer from the ensuing austerity without the Barnett formula and the other perquisites of the British state, a tactical vote for a Tory is very definitely the lesser of two evils and necessary for the greater good

It is a measure of the depths to which identity politics have sunk in Scotland that trolls from all parts of the nationalist community and some in the Labour community have crawled out from underneath their coffin lids to denounce him. George has compounded the heinous sin of voting Tory ('How could you?!') by teaming up with that most satanic of creatures, an Etonian. They have delighted in digging up George's past announcements, 'If you ever see me standing under a Union Jack shoulder-to-shoulder with a Conservative, please shoot me.'

It should not have come as any surprise – though evidently it has – because, we have made it very clear that our modus operandi is to break the depressing stalemate in Scottish politics and defeat the SNP by making tactical voting work. And the sitting MSP in George's Dumfries constituency, with the best chance of winning against the separatists, just happens to be a Tory. George, with his canny political nous and his 360,000-plus Twitter followers, has just publicised the strategy and the cybernats have helped to amplify the message. Even the justice secretary Humza Yousaf couldn't resist putting the boot in (while taking care not to tag George in his tweet).

Tories are less vocal in dealing with perceived traitors but I have also had aggressive emails from tribal Conservatives accusing me of 'splitting the vote' by daring to suggest tactical voting. It is becoming increasingly obvious that very few Scottish voters understand the voting system we have been saddled with since devolution, and equally clear that the big parties would like to keep it that way. For them the process of politics is more important than the outcome of the election. It is more critical for their party leaders to maintain their share of the vote than it is to evict Nicola Sturgeon on 6 May. It is also impossible for them to keep their party machine motivated and intact if they endorse tactical voting, which produces losers as well as winners among their candidates. So it suits them to trot out their message: 'both votes to the Tories/Labour is the best way to defeat the SNP', knowing full well that this is simply untrue. The d'Hondt method of allocating the second 'List' votes deliberately favours smaller parties that have not won seats with first votes in the first-past-the-post constituency battles. Thus if, with George's help, the Tories win four seats in the South of Scotland, their votes will be divided by five (one plus number of seats won) in the List. This means that tactical voters have more chance of securing their desired outcome of defeating the nationalists by giving their first vote to the best placed 'pro-unity' candidate and their second vote to All for Unity. In fact, giving a second vote to one of the big parties would be to reduce its effectiveness. If All for Unity can translate those votes into more anti-nationalist seats then it is game over for the SNP and we will be in a position to secure a pro-UK coalition in Holyrood.

* * *

The SNP is imploding nicely as the Salmond Inquiry, more properly the Sturgeon Inquiry, continues to shock high-minded separatists like Jim Sillars into speaking out against the rampant corruption in their party. Although the polls still show an SNP majority with the Holyrood election in May only weeks away, and even our state broadcaster appears to be backing Sturgeon's party, I really think we can win this. I just wish the Scottish Conservative and Unionist Party thought the same.

It is a confusing time to be a Conservative in Scotland. The Scottish Tories have struggled since devolution to cope with the accusation that they are a 'branch office' of the national party. The taint is reinforced by the unfortunate habit of the more capable Scottish Tories – Michael Gove, Alister Jack, Ben Wallace – heading off to Westminster. One of them should really be leading the Scottish Tories into the lists in May and not Douglas Ross, who has many qualities but is doing the job about fifteen years ahead of his time – and had, in any case, decamped to Westminster himself, whence he struggles manfully to lead the party by remote control. I supported Murdo Fraser's bid for the leadership in 2011 precisely because he proposed to reverse Edward Heath's misguided reforms and re-establish the Scottish party as a completely separate entity, though calling it the Scottish Unionist Party would have been a mistake for sectarian reasons. It would really be much better if the parties in Scottish domestic politics were separate from the ones contesting national elections, something that happens in some federal countries.

The confusion is compounded by the Scottish Tories' perhaps understandable desire to distance themselves from the national party by moving ever leftwards. It is often heard said that there is no centre ground in Scottish politics, but Douglas Ross has gone to great lengths to find it. He and his party were 'remoaners' long after it was fashionable (and I write that as a former Remain voter myself). He has also backed the SNP's very un-Tory policy of free university education for all, ironically a policy that has actually been shown to militate against poorer students and reduce social mobility.[57] More recently he has voted against the government on internal market legislation. So we are faced with a situation where pro-UK Scottish voters have a choice between voting for a party led by a privately educated millionaire businessman, the Labour leader Anas Sarwar, who faced down his party to send his own children to private school ... and a party led by Douglas Ross, a state-educated former farm worker. The overwhelming impression is that they are two social democrats dancing on a pinhead, obsessed, in George Galloway's memorable phrase,[58] with 'the narcissism of small differences'.

At the moment it appears as though all the Scottish Tories want out of May's elections is to have their share of the vote held up and to come an honourable second. Douglas Ross said as much in his first announcement on becoming leader when he said that he wanted to be the 'best leader of the opposition' – a gaffe that was then hurriedly corrected.

* * *

57. This is because it has led to a quota of university places for budgetary reasons.

58. Paraphrasing Sigmund Freud.

'See if you can reach that one going under the slates.'

I am up a ladder where I have been each March since not long after we clothed the front of the house with that loveliest and most invasive of climbers, the wisteria. The Chief Horticulturalist stands at the bottom, seemingly unconcerned about the imminent risk of widowhood, as I lift my head over the fascia so that my chin rests in the gutter and gingerly reach across the roof to find the offending tendril and give it a yank. As I do so I reflect on how the characterisation of Adam and Eve makes the story of man's original fall in the Book of Genesis so believable. There is a serpentine quality to *Wisteria sinensis,* China's finest botanical export. I love the way it winds itself around drainpipes with oriental wiliness until I have to go and free them before it threatens to prise them from the masonry. On my ascent I passed last year's goldfinch nest in a thicket of branches where the fuchsia, the rose and the wisteria meet in a pleasing composition. And higher still the muddy outline of an old swallow's nest reminds me that our colony must soon be leaving their winter hideaway in the South African veldt to come home, if they haven't already. The wisteria is studded with leaf buds and is also carrying a good number of flower buds this year. Pruning in the raw cold of an easterly wind makes me marvel at the way the change in seasons will transform the façade of the house into a pastiche of an Oxford college in a matter of weeks with bright green foliage and blue cascades of wisteria flowers setting off the honey-coloured stonework. The promise of it makes me hold on to the ladder more tightly.

Most of the cows have calved now and are being milked from grass. 'The lowing herd wind slowly o'er the lea', as Thomas Gray put it, after each milking. We are also adjusting to our annual influx of walkers, swelled by the lockdown, who have so far co-existed without complaint on the cow tracks with cows, their excretions and mains electric fences. There appears to have been a corresponding increase in litter. The ladies of Dumfriesshire, led by my friend the artist Minette Bell Macdonald, have responded with a heartening Covid initiative by meeting regularly to walk together patrolling the verges and lay-bys and picking it up. Candidates for 'lockdown heroes' are surely the ones who empty the bottles dumped by incontinent lorry drivers.

* * *

The county normally comes together at this time of year for funerals and the inability to hold a fitting send-off will be an abiding sadness for many families. But some lockdown funerals will be remembered with affection. I attended one recently for a family friend in her garden. Celia's coffin was placed by the bird table outside the conservatory where she had spent her dotage. We were all told to bring warm clothes and garden chairs. The old and disabled parked their cars up close so that they could watch through their windscreens. It seemed the most natural thing in the world to say goodbye to someone in the garden she had lovingly tended and where I had played as a child. As we left, lots of mourners were heard to say they would like the same. I predict that funerals, like weddings, may not all go back to church after the pandemic.

* * *

Like Bob Geldof and the Boomtown Rats, I don't like Mondays but today has been more depressing than usual. The first setback was hearing the BBC describe Nicola Sturgeon being cleared by an 'independent' inquiry. The second was reading an op-ed in a newspaper by the *Spectator*'s Scottish editor describing the SNP as 'exhausted but irreplaceable'.

The riposte to the first has to be that if this was an inquiry into Donald Trump or Vladimir Putin, or almost anyone else, the BBC would not use the word 'independent' about an inquiry by an inquisitor carefully selected by the accused and limited by very delicately confined terms of reference before delivering a report that is at odds even with the leaked conclusions of the Parliamentary Inquiry, which is then so heavily redacted as to be meaningless. Banana republics are most egregiously defined by the complicity of their state broadcasters and BBC Scotland appears to be no exception. Sadly.

The second is perhaps more worrying. If even a nominally unionist senior journalist believes that May's Holyrood elections are such a shoo-in for the SNP that it is pointless for the Scottish Tories and Labour to fight on a manifesto of change, then it is time to start packing and follow the steadily growing convoy heading south.

It begs a question. If now is not the time to replace the SNP, when even nationalists like Jim Sillars and Craig Murray and the Rev. Stu Campbell say they won't vote for them, then when will it be? The gangrenous narrative that would accept another SNP-led coalition at Holyrood, while London plays a long game and triumphs over separatism at some undefined point in the future, is deluded and a betrayal. It is deluded because it would mean another five years of brainwashed young people exiting our schools and universities with a *Braveheart* education. And it is a betrayal of parents and patients and drug addicts and commuters and just about anyone outside the complacent Holyrood bubble.

CHAPTER 14

Campaigning

George Galloway and I launch our campaign by filming a 'Potemkin rally'[59] in the pub car park – all that we can do during lockdown.

We are met by the village loudmouth, ostentatiously masked by a black scarf against the Covid on the cold March breeze. 'What's going on here? Did you no see the Saltires as you drove in? We don't want you and your union here.' He patrols the one street in the village while we are filming, just in case any 'Yoons' think about joining in.

Despite this several older people do come and quietly give us words of encouragement. 'We'll be voting for you. Good luck.'

Later I speak to a retired police sergeant, who comes up to greet me and meet George. He is an ex-soldier. I have him down as a floating voter and probable unionist. He and George blether about football. George has an encyclopaedic knowledge of the game and had trials for several Scottish clubs in his youth.

Then the bombshell:

'I'm sorry but I'm going to vote SNP this time. I want independence. I want to do it for my boy here.'

George looks astonished. 'Seriously? After all the failures and the sleaze and corruption that's coming out? You still trust them?'

---

59. Staged as if it is a real rally but without people, as Covid restrictions don't allow gatherings. The speeches are then broadcast on social media. It's the best we can do. It really isn't a 'free and fair' election under Covid restrictions as it favours the established parties.

'Och they are just as bad in Westminster. What about the corruption down there?'

The standard excuse, how depressing. No, they really, really aren't.

He then proceeds to tell us, without any doubt or irony, how Scotland subsidises the rest of the UK, something that has apparently been suppressed since Denis Healey took the decision to classify the exact amount back in the 1970s. And, without the dead weight of the rest of the UK, Scotland will definitely have a much brighter future, especially when we get back into Europe, he opines.

Our amazed rejoinders about the black and white evidence of the GERS figures showing Scottish reliance on the Barnett formula fall on deaf ears. He doesn't buy the near impossibility of getting back into the EU either. As Mark Twain said, 'It is easier to fool people than to convince them they have been fooled.'

And there you have the nationalist formula ladies and gentlemen: intimidation on the streets ... the Big Lie Theory, patented in Nazi Germany and still being put to good use by the SNP ... sprinkled with 'whitabootery' – the Teflon coating that smothers any criticism of the Great Helmswoman.

Next up is Ayr. As we cast our words into the salt air of the Irish Sea on the beach front for the benefit of passers-by, and journalists from the *Scotsman* and *Daily Mail*, two single vapour trails cross in a clear blue sky behind us in the shape of the Saltire.

We appear to be well received by the townsfolk, then our security people draw my attention to a large 4x4 across the road. It has two men wearing sunglasses in it and SNP stickers in the windscreen. One of them is talking into a phone, the other stares at us fixedly.

Shen, our security man nods. 'The opposition have turned up. They have been watching us for a while.'

Suddenly I am back on patrol in Northern Ireland. There we would frequently be subjected to 'dicking' where IRA sympathisers would satellite patrols, reporting our whereabouts to the terrorists. Often it was done overtly and deliberately to intimidate young soldiers.

As we come to an end, Police Scotland turn up. They have had an illegal gathering reported to them. They are satisfied by our explanation and we go on our way. But it leaves a bad taste.

The more we get into this election the more we realise what our country has become and the more our resolve hardens.

* * *

Last week started with a YouGov poll finding that George is the best-known opposition politician in Scotland and the one that voters in several regions think would provide the strongest opposition to the SNP. We use that on the soapbox where George announces that he will be 'holding his nose and voting Tory' with his first vote.

If Scottish Tories were grateful for the endorsement of Scotland's best-known left-winger, they had a funny way of showing it. Disquiet in the Tory camp that George is better known than their leader, Douglas Ross, led to aggressive trolling by anonymous Tory accounts and knocking copy from journalists seemingly primed with the narrative that All for Unity is a 'fringe party that won't win seats and is only going to split the vote'. One of the first rules of politics is to ignore your opponents if they are smaller than you, so that you don't give them publicity. It is strange then that, in one of the most momentous weeks in Scottish politics, when every sinew should have been strained to topple Sturgeon, so many memes, tweets and column inches have been devoted to attacking us. The notion that the big parties have some sort of loser's entitlement to seats on the list after failing to win constituency seats – because they in turn have split the pro-UK vote three ways – is the politics of the absurd.

The week has ended shockingly with George and his young family having to move out of their home after death threats from a nationalist.[60] Proof that if some journalists don't see All for Unity as a threat to the SNP, the separatists do.

* * *

---

60. The man was swiftly caught, charged, and later found guilty and sentenced to community service.

Journalists attack our lack of experience as politicians. We are delighted to fight on this turf. Arthur Keith is the only veterans' spokesman of any Scottish party to have heard a shot fired in action, as ex-RSM of the Black Watch, and the same goes across the piece. Our Justice team is court lawyer (barrister) Charlotte Morley, former Procurator Fiscal Moira Ramage and former tribunal judge Christian McNeill. Our health spokesman Dr Bruce Halliday is a GP with twenty years' experience in the NHS, as is Dr Jon Stanley our social care spokesman; our education spokeswoman Linda Holt was a lecturer at Oxford University, and so it goes on. Several of our candidates are best-selling authors. I don't want to be rude but we are being adversely compared to parties led by career politicians who have never held down a real job. Seriously?

The truth is we are still small. We have come from nowhere, as all parties once did. The Electoral Commission did its best to strangle us with delay and bureaucracy. BBC Scotland is still ignoring us. And now the establishment seeks to erect barriers to entry. That is politics. Indeed there are probably some voters who have yet to hear about All for Unity, rather fewer after George Galloway's performance in the *Scotsman* hustings for the South of Scotland this week that left rivals gasping like goldfishes on the carpet. By common consent he won the debate hands down, although I may be biased in that opinion. Certainly the newspaper devoted most of their report to him when they wrote it up. The other candidates spent most of the debate attacking George. The Tory candidate accused All for Unity of splitting the vote. Then, when George retorted that he was himself helping the SNP by standing in East Lothian, which is a winnable Labour seat, the Tory smoothly quoted statistics from the recent Westminster election showing that Labour and the Tories are neck and neck. They aren't. That is a different constituency with different boundaries. The equivalent of a grocer responding to a complaint about his apples with a response about pears. It's no wonder people regard politicians as untrustworthy. Afterwards George accurately describes the Tory candidate to me as a 'garagiste'.

* * *

'A week is a long time in politics' is a trite cliché but it may have been true this week as Alex Salmond's forced entry into the race has left things even more uncertain. It is what we have feared all along: the formation of a 'Nationalist Front'. It makes last week's squabbles with other parties accusing All 4 Unity of 'splitting the vote' seem rather petty now.

Salmond's new Alba Party has been greeted by some as good news for unionists as it will put Salmond and Sturgeon's Macbeth-like struggle centre stage in the election and it will 'split the vote'. Whilst the former may be true, the latter argument does not bear scrutiny from anyone who understands Scotland's Byzantine electoral system. The truth is that there is now a list-only separatist party capable of converting previously impotent nationalist second votes into seats and therefore into a super-majority of MSPs for independence (but crucially *probably* not voters). The separatist monster now has two heads and will therefore be harder to slay.

It helps us a bit as we are the mirror image of Alba and voters are starting to look more closely at the electoral system. Like Alba, we are not standing in constituencies and can therefore maximise the pro-UK vote. Salmond's intervention makes it even more critical that the unionist side also games the system.

***

Our candidates find themselves being trolled aggressively by angry candidates from the main parties producing ever more fanciful charts to show hypothetical situations where A4U replaced their candidates in list seats. They all seemed oblivious to the fact that if it came down to the main parties worrying about list seats we would have lost anyway and be off to Catalonia with Sturgeon and Salmond. If anything it has made us even more determined to get the electorate to go rosette colour-blind and vote tactically.

***

As the A4U team works together it is becoming more and more apparent that there is far more that unites us than divides us. We come from nearly every party in Scotland yet there has been no hint of any of the disagreements we had anticipated. George and I were a little wary of each other to start with. He has been battling all his political life on a different side to me on most issues, and perhaps we will again one day. But I am sure we will remain friends and we surprise each other daily by how much we agree on. It is abundantly clear that the real fault lines in Scottish politics are completely different now. They are now unity versus separation, libertarianism versus authoritarianism, Enlightenment values of truth and reason versus post-modernist post-truth, the old politics of tolerance versus 'woke', free speech versus 'hate crime', transparency versus corrupt cronyism.

If George and I can agree it makes the animosity between Sarwar and Ross seem ridiculous. It is like one of those sociological experiments where a group of normal people are given different rosettes and put into groups and are then observed to start hating each other. Politics of the absurd again.

I put this hypothesis to one of my lefty friends and he tells me that anyone growing up in the Central Belt during the Thatcher era will never ever vote Tory. In his street families went from having their fathers in jobs to having unemployed dads in an instant and many have not worked since. They don't blame the Tories for closing the mines, the shipyards and the steelworks, they knew they had to go anyway. What they will never forgive them for is not having a Plan B and leaving them on the dole for a generation. It is a depressing insight as it explains why tactical voting is a hard sell.

* * *

The standard attack against All for Unity now is that we 'don't exist outside Twitter'. If that is so, why bother to attack us? However, there is some truth in this and the big challenge is to get our message out there through other media. But when you analyse what Twitter is you come to an uncomfortable conclusion. Twitter is peopled by journalists, and George, I and the A4U Twitter feed are followed by nearly every newspaper and TV editor in Scotland. Some quote our tweets, which are the modern equivalent of press releases, many still do not. Could there be sections of the media who don't want us to succeed, either because they are pro-independence or because, like the political parties they support, they too are tribal? It's a chicken and egg situation.

***

Our suckler herd is now fading into memory. And the patchwork of stubble, plough and grass we once had at this time of year, with our old mixed beef and arable operation, has given way to a sea of grass. I would never have imagined just how much grass until we adopted the new mob grazing system. Some fields have been grazed twice already and spring has barely begun. We are grazing the cows in tight paddocks and so they look, from a distance, like a herd of wildebeest on the Serengeti. The queue for milking snakes for half a mile across the landscape and can lead to a traffic jam of two or even three cars at the road crossings. How do they manage in suburban parts of the country?!

***

We are facing change at the kennels where our master and huntsman for the last fifteen years, Andrew Cook, is retiring on 1 May. Andrew has the distinction of being the first huntsman of a pack of hounds founded after the law was changed. To the largely unappreciated and un-thanked travails of being an amateur huntsman, the risk of arrest and even imprisonment for the misdemeanours of thirty canine vulpiphobes over several square miles has been added. It is hard enough for humans to understand the absurdities and inconsistencies of hunting law let alone for illiterate foxhounds. Andrew has borne these constant anxieties with great coolness. It is largely thanks to his enthusiasm that some derelict farm buildings are now home to five horses and twenty-five couple of outstanding hill hounds. More importantly, under his stewardship the kennels have become the hub for a community that stretches for thirty miles or more in three directions (we are bounded by the sea to the south!). Wayward children have found something that enthrals them, lonely old people have found a way of spending winter afternoons in the company of others;

and new friendships have been forged across conventional social boundaries. It has been a bitter disappointment that the Covid pandemic curtailed Andrew's last season but he leaves behind a committed band of enthusiasts who will be raring to go as soon as autumn hunting starts again.

I am so grateful for all the good times we have had hunting together. Living cheek by jowl with the kennels I have watched Andrew up to his elbow in foul drains, bicycling miles with his hounds to get them fit on early summer mornings, or arriving back late at night after hunting after standing for hours in the dark on a windswept hillside gathering hounds out of a forestry block long after the field had gone home. Huntsmen are not allowed to take the day off during the hunting season, or to be ill. The only time he was 'off games' was when he broke his arm when he was kicked by a horse and Ben, our whipper-in had to take over.

Our new master George Humfrey is a lucky man.

* * *

Attempts to unify the unionist vote in the Scottish elections appear now to have failed as the Scottish Tories and Labour have brushed off our appeals to work together and the reality is that with the major parties splitting the anti-nationalist vote in constituencies again, and Alba having the capability, if not the certainty, of converting previously impotent second separatist votes into seats on the AV system lists, there has to be worst-case contingency planning for the nationalist 'super-majority' that Salmond craves.

Those of us with direct experience of the Troubles in Northern Ireland have been warning for some time about the 'Ulsterisation' of Scotland and the slide towards a 'Scottish troubles'.

This putative super-majority of seats would be used to attempt to ramrod through secession. The outlines of a more hard-line nationalist strategy have been evident for some time after Joanna Cherry's proposal to copy de Valera and 'start negotiations straight away' without bothering with another referendum. There is no chance of the separatists gaining enough support for copying the unilateral declaration of the Irish Free State, which resulted in a bloody civil war, not least because Westminster would simply choke off the money supply. But it was the threat that counted. And now Salmond's implied threats of civil disobedience should be taken seriously. Salmond even quoted the Irish nationalist leader Charles Stewart Parnell in his launch speech. It has to be remembered that the recent Troubles in Northern Ireland started when civil rights marches spiralled out of control. But in Ulster the police were under direct control of a unionist civil power and resolutely British, whereas in Scotland the newly centralised Police Scotland is under the direct control of a Scottish government that would, in that scenario, be separatist.

This muscular nationalism also assumes, let's hope wrongly, that condemnation of a Catalan style revolt from outside the UK may be more equivocal now that we are outside the EU and there is a Biden Presidency with a worrying history of sympathy for the republican cause in Ulster.

I wish unionists would stop having a sterile argument amongst themselves about vote splitting in the lists. If there is a nationalist majority in Holyrood, the numbers of seats individual parties hold would be far less relevant than the overall number of votes cast for pro-UK parties. It is vital to get the pro-UK vote out on 6 May and to be able to demonstrate a silent majority for keeping the United Kingdom together. We need to move the focus away from seats won and onto the split of votes between separatist and pro-British parties. Commentators, especially the BBC, need to be careful not to conflate a majority of seats won under an absurd electoral system and a 'mandate' for secession. And we should shine a very bright light on any attempts at 'Ulsterisation' and condemn it robustly. And pray.

* * *

George Galloway and I would never have imagined that we would one day be standing on a platform with each other. It has led to a mutual understanding. Today, at my suggestion, he addressed the virtual Rural Workers' Protest online, organised by the Scottish Gamekeepers' Association and spoke, as a left-wing trade unionist, of the 'tyranny' imposed on the countryside by unthinking regulations. Now that is different!

***

They say, with penetrating accuracy, that once you become a parent you are only as happy as your least happy child. And, sad to say, since we got the Alliance for Unity going last year it has been like having an extra child. The party's triumphs and disasters have produced a similar reaction in me as my children's and I find myself worrying like a parent about our progress. The anxiety initially was not finding enough people of sufficient calibre to join us; then it was that the Electoral Commission would find excuses to turn down our application to be a political party until beyond polling day; then it was that we would simply be ignored by everyone; then it was that we would never feature in any polls. So the *Sunday Times* poll putting A4U on 4 per cent, 1 per cent behind the Liberal Democrats, is a bit like seeing one of the children winning their first egg and spoon race, It is heartening and frustrating at the same time. Heartening that, after weeks of being lumped under 'other' in the question and therefore being shown under 'other' in the poll result, we finally exist on Planet Psephology. There is intense frustration that had George or I been included in the BBC Leaders' Debate we would surely have polled much higher. There are still voters who have never heard of us thanks to the media's attempts to ignore us. With more publicity we can go much higher.

***

Our *Sunday Times* poll debut appears to have triggered a nervous breakdown in the Tory Party. I had resolved to spend the Easter weekend having some family time but end up fighting a defensive battle against Tory trolling on Twitter. I find this a bit odd as I have been a member of the Tory Party much longer than Douglas Ross (who traded up from the Scottish Liberal Democrats and was born the year I first voted Tory) and I suspect my views are more in line with orthodox one-nation Toryism than his. Tory Party HQ seem to have lined up Jackson Carlaw and some MSP called Maurice Golden to mount personal attacks on George. It looks as if they are also using the journalists Stephen Daisley and Henry Hill to stir up animosity towards A4U. And I detect plotting with the supposedly non-partisan unionist lobby group Scotland in Union, who are busily telling anyone who still listens to them not to vote for 'fringe or new parties'. Scotland in Union seems to exist in order to exist. They suck thousands out of unwitting unionist donors each year. Most of which seems to go on high running expenses. Certainly there is very little discernible output and they are completely overshadowed by much more effective volunteer groups such as TheMajority.Scot and Scotland Matters.

Dear old Tories, they are completely blind to the irony that in running total no-hoper candidates in what should be safe Labour or Liberal seats they are doing far more to 'split the vote' than we ever will. And George is not exactly a Tory icon coming to steal their votes.

* * *

We arrive at the studio in Glasgow for the manifesto launch. George, being the old media pro that he is, immediately starts to 'have his face put on' by Gayatri. I tell him it looks as if he is being embalmed. He agrees that may not be too far wide of the mark. I have opted to wear a tweed coat and my regimental tie. I only ever wear a suit for funerals these days and I have made a mental note that if I am ever seen in a dark suit, white shirt, blue tie, Bob the Builder hat and high-vis jacket, everyone has permission to shoot me. Nevertheless I feel rather a country bumpkin in the middle of Glasgow. I have decided to go against my normal modus operandi and prepare a speech to read off the autocue. When we get there the autocue is nowhere to be seen so someone holds it on a laptop next to the camera, which makes me look as if I have a bad squint. The press conference seems to go okay, although the BBC only seem interested in gotcha journalism, trying to trip George up over his appearances on Russia Today. It makes me wonder why we bothered to write a manifesto.[61] It's an enjoyable experience though and Ricky and his team in the reflexblue studio have really pulled out all the stops and provide mugs for our coffee with the A4U roundel symbol on.

* * *

The manifesto launch finally gets us onto the BBC. George gets an invitation to be interviewed by Martin Geissler on 'the Nine' on BBC Scotland. Geissler appears to have been set up by the producer to do a demolition job on George, bizarrely accusing him of standing alongside 'Old Etonians, former army majors ...' (that'll be me then). It fails and I suspect from now on Geissler will have his nightmares peppered with the words 'You're not Jeremy Paxman, trust me on that.'

* * *

'Fog in the Solway, the United Kingdom cut off.' I paraphrase that famous *Times* headline about the Continent in my head now every time I lose sight of the English coastline just nine miles away across the Firth from our kitchen window. There are now three and a half weeks to go before the election, which may be the last when Scots vote as British citizens. It lends a surreal quality to this spring. Daffodils, primroses and cuckoo flowers put on their shows in all the familiar places. Bluebells and garlic flowers are about to pop in the woods and we search the skies as usual for the first swallows. Calves and lambs concentrate minds on every farm and all eyes are on the grass growth in the dairy paddocks as it swells with each hour of warmth or checks in the late frosts. Yet despite the outward normality there is an unspoken fear that, although the natural world will carry on as normal, Scotland may be about to

---

61. Despite the broad spectrum of our candidates from Left to Right we all agreed on a manifesto to slim down Holyrood and devolve power to the regions, park the independence issue through a Clarity Act, reform devolution and clean up public life in Scotland by bringing in constitutional safeguards, de-politicise education, and de-centralise the police inter alia.

be 'changed, changed utterly' as W. B. Yeats wrote about Ireland in a similar context a century ago.

* * *

Since we split our farming operation into two halves, the dairy takes up the north end, which is now one big green park through the seasons. The south end remains in an arable rotation and April is its turn to jump into life after six months of brown stubbles and solitude. The middens steam like power stations on frosty mornings as they are opened up for the muck to be spread. And contractors crawl over the land like large sci-fi insects with every manner of contraption to till the soil for barley and potatoes to go in. The appearance of seed on the ground again after the lean months has attracted every pigeon, jackdaw and rook in the district. Jackdaws are notorious egg thieves and pose a real risk to songbirds. I have more sympathy for rooks, which will eat eggs if they find them, but otherwise do a grand job of clearing our fields of slugs and leatherjackets. I grew up under a rookery and spring would not be the same without their raucous toing and froing in their treetop building sites. By strange coincidence, the year the big house was sold the rooks left as well and took up residence above the school playground in the village. I like to think of another generation of children being captivated by their noisy squabbles.

I take an afternoon off to shoot the pigeons coming into the seed fields over decoys with Oliver. Sitting quietly on a warm spring afternoon behind a lacy screen of blackthorn blossom on the edge of a wood is a welcome respite from the vicissitudes of the political metaverse. The only sound comes from the first bumble bees foraging on the spring flowers and I am rewarded by seeing a weasel making its way inquisitively along the wall two feet in front of me. Some things will never change.

* * *

'We'll provide free bikes for all children of school age who can't afford them ... #BothVotesSNP.' Yes, you read that right. No need for Scots to buy their children birthday presents this year, the government will do it for them. And if they already have bikes, they can always shove them on eBay. Winner! This egregious vote-buying follows on from promises to buy every schoolchild a laptop, a £20 per week child payment, free school meals for all ... The final insult is the SNP's announcement that Scottish tax rates will be frozen.

You have to admire their brazen chutzpah. They know their core vote is people under forty with school age children. They also know that likely SNP voters believe the Big Lie: that Scotland puts more into Her Majesty's Treasury than she gets out. So that insulates them nicely from the obvious question: 'How on earth would we afford all this without the Barnett formula?' And the more fury it induces among taxpayers in the rest of the UK, who know that the Big Lie is a lie, the better. It feeds the Little Englander narrative that says, 'We'd be better off without them whinging Jocks. Let them go, they'll be bust within a year.' Those who scoff at the 'Champagne Separatist' Ian Blackford, as he huffs and puffs and flaunts his telephone number expense claims, haven't been paying attention to a highly skilful political operation to irritate the English taxpayer. If Boris caves and grants another referendum it will be partly because of a growing antipathy towards Scotland south of the border.

If you think that this is state sponsored corruption then that's because it is. All governments bribe the electorate with their own money to some extent; it takes a special sort of dishonesty to bribe voters with other people's money. The strategy has been evident since the inception of the Covid business support grants last year. The SNP appears to have been carefully siphoning off some of Rishi's millions into a pre-election war chest. As a farmer who relies on self-catering holiday cottages to underpin the vagaries of agriculture, I have special reason to be aggrieved. While our competitors in the Lake District were receiving £10,000 per empty cottage, Scottish ones only received £7,500. We saw where some of the money was going when Scottish NHS workers received a tax-free pre-election bung of £500 each as a thank you for their service during Covid. Wealthy dermatology consultants, who had little to do during Covid, received a bonus, while ironically many care home workers did not. But in the cynical world of Scottish politics that didn't matter. The SNP had decided that, while Sturgeon was portraying herself as the saviour of Scotland on television, health workers would be another part of their core vote and they got their headline.

If the polls are accurate, for every demoralised business owner, few of whom would ever vote SNP anyway, there appear to be more voters who are prepared to take the money and not ask too many questions. The unionist parties are still splitting votes in the constituencies. The Labour and Conservative leaders are fighting over who will come second – Anas Sarwar's posters even argue that he will be the 'best opposition'. Only mass tactical voting, as we are arguing, can save Scotland from another five years of corruption now.

* * *

I decide to do my bit for the A4U television party political broadcast[62] in a field with our dairy cows in the background. The sun is shining and by chance they are grazing one of the paddocks that has Criffel, the South of Scotland's most beautiful hill in the background. George's wife Gayatri is behind the camera and, despite being well outside her comfort zone, gamely scrambles under the electric fence and gingerly picks her way in patent leather ankle boots between the cowpats to set up. It takes several takes, as each time I launch into my soliloquy the cows come up and start nuzzling me so that we get the giggles. Gayatri clearly thinks I must be St Francis of Assisi. I take the opportunity to make another video explaining how vulnerable the dairy industry is to a hard Scexit. There is nowhere near enough processing capacity to deal with all the milk produced in Scotland. Thousands of litres of milk go south as far as the English Midlands each day. A hard border at Gretna would be a disaster. Farmers are also terrified of the SNP's currency

---

62. Under Electoral Commission rules we are allowed one free TV broadcast because we have a full slate of fifty-six candidates on the List. We make it ourselves to save money.

plans (or lack of them) as we all have large loans in sterling, which we would be hard pushed to service in any new, devalued currency.

* * *

The trolling by other unionists continues apace on Twitter. One minute they accuse us of being a threat to their parties, the next minute they say we are not going to get any votes anyway. Hmm! It's as if Tesco, Morrisons and Asda are telling consumers not to shop at Aldi as it might depress their market share. There are actually thirteen pro-UK parties standing on the list. If All for Unity had not launched a charm offensive last year to try to get all unionists under one banner, there would be seventeen because the Scottish Unionist Party, Workers Party, Reclaim and the Scottish Christian Party would all be contesting this election separately, but we are never given any credit for that. We tried very hard to get the leaders of four other small parties to subordinate their egos for the greater good but sadly they declined. They know who they are.

I now understand why previously friendly Tories on Twitter are attacking us so aggressively. A Tory, disgusted by the party's arrogant sense of entitlement to list seats, has sent me a leaked document entitled *Rolling Brief: Scottish Parliament Election 2021* produced by the Scottish Conservative Research Department (I wish we had one of those). It tells party members to accuse us of 'splitting the unionist vote' and encourages ad hominem attacks on George for his previous speeches in support of a united Ireland. I think most voters are sophisticated enough to understand that Ireland and Scotland are very different. Nevertheless it makes us a target for the more extreme type of unionist, which is a frustration for A4U candidates like me who risked our lives to keep the peace in Northern Ireland, ironically while the young journalists attacking us were still at primary school. All for Unity is standing fifty-six list candidates in this election and we have published a manifesto of the ideas we wish to implement to make Scotland a better place. Yet many sections of the media, and especially the BBC, only want to focus on one candidate: George. I suppose it's inevitable but it is frustrating.

The leaker's assumption is that the Tories' internal polling is showing All for Unity doing far better than they expected. They would rather have ignored us. There is a rather naïve assumption among their more unquestioning followers that A4U will simply evaporate if they keep attacking us. They forget that if they were to be successful in keeping A4U below the level at which we would win seats it wouldn't help the unionist cause. It would be better to get behind us so that our votes, which are not divided by any d'Hondt handicap, are leveraged to the maximum. They also forget that for every one person in Scotland who supports the Tories there are four who don't, so making A4U their enemy risks making lots of other voters our friends.

Their 'both votes Tory' slogan flies in the face of every bit of tactical voting advice. It seems to be rather like circling the wagons and hoping the Apaches have run out of arrows. I despair of my parent party sometimes.

I wonder briefly whether we should pull out of the race. We are still not breaking through as I would like and the charge that we are going to harm the unionist cause by running needs careful assessment. But it is clear that whatever happens the SNP is going to win enough seats to form a government, with or without other separatist parties, so the challenge is to get as many unionist voters out on the day so that there are more pro-British votes in ballot boxes – and we have an important role to play in achieving that.

* * *

As the campaign progresses I am more and more impressed by our candidates. It has been a risky business selecting fifty-six people with whom to stand shoulder to shoulder when we were not allowed to interview face to face because of Covid, particularly as unlike any other party we have gone out of our way to try and represent just about every viewpoint as long as it isn't nationalist. It has been astonishing to see how socialists and capitalists, Protestants and Catholics, townies and countrymen have all pulled together. We have resisted the temptation to be too prescriptive in telling our candidates what to say and this has led to the flowering of some great talents on YouTube videos – Jean-Anne Mitchell the feisty Glaswegian granny and ex-Labour candidate on the SNP's education policies, Philosopher Dr Catherine McCall's impassioned defences of free speech, Niall Fraser's oratorical brilliance on the subject of SNP perversions, Charlotte Morley like a tigress in defence of her cubs, and many more. I feel sure that whatever the result of the election Scotland will not have seen the last of these people.

***

The grey geese left without saying goodbye and the only evidence of their stay is a growing pile of scarecrows in a corner of the old stone barn – high-vis yellow jackets on wooden crucifixes, which seem to have kept them off the new grass leys quite effectively though there are other signs: goose droppings and bare patches in certain favoured fields. The barnacle geese are still around but in fewer numbers and concentrated near the golf course. They seem restless. Taking off for no reason and circling round before landing again, as if in training for the long flight back to Svalbard. I wish that I could report a balancing influx of swallows but they are late this year. I saw my first ones on 29 April, hawking around the cattle – always the best place for insects. I hope it is just that they have been held up by the cold northerly winds but there is no escaping the fact that there are fewer every year.

This long, hesitant spring continues to frustrate. I have grown up with the reasonable expectation that it should rain in April. It is the month of showers in weather lore. Well, not here, not this year. I have been chronicling the phenomenon of back-to-front weather patterns for some time – dry springs and early summers followed by monsoons just as harvests start. The year 2021 seems to be shaping up to the new normal. Those who want to find good reason to close down the rural economy will no doubt evidence it as climate change and blame it on belching cows or grouse shooting. One consolation is that we did not sow any winter crops last back end; they would be looking very sickly if we had. The new dairy has so far managed to cope with the drought. Joey benchmarks what we are doing against other dairy farms in the area and I detected a note of *schadenfreude* when he said that all the others were having to feed their cows silage to keep the milk yield up. There are upsides to our heavy land, which retains the moisture better than most. The trees have sensibly delayed coming into leaf until the weather changes. It has been the best year that we can remember for primroses. The lack of rain to spoil their blooms has kept them in good shape for longer than usual and there are many more clumps this year – 2020 must have been a good year for them seeding.

The long lockdown has come to an end at last. This has led to a frantic rush to get holiday cottages ready. The kitchen table is covered in rolls of material and the house rattles to the sound of the sewing machine as the Chief Interior Decorator puts the finishing touches to new curtains. Paint pots are hurriedly deployed to walls that have developed mould spots in cottages that have lain empty for months and there are last-minute searches for light bulbs and smoke-alarm batteries. It is good to be back in business but one part of us is secretly annoyed that our peace is about to be disturbed again. One legacy of lockdown is the bigger than usual pheasant population for the time of year, a reminder of our Covid-curtailed shooting season. No other birds have such a visible natural selection process and there are cocks fighting everywhere or strutting magnificently with their harems while their rivals eye them from the fringes. It is a miniaturised version of red deer during the rut.

* * *

'You know there was no word for dildo in Gaelic[63] until the SNP invented one? They have paid translators to come up with the full suite of words to cater for all sexual tastes.'

The breadth of George's grasp of current affairs never ceases to amaze me. We are blethering in a café in Stranraer on our soapbox tour of the South of Scotland. This conflation of the SNP's fake Gaelicisation, warped morality and waste of taxpayers' money would have been a gift for opposition politicians and the media in most countries. The fact that it has barely permeated the public

---

63. Just in case you are wondering, the Gaelic for sex toy is dèideag gnè.

consciousness in Scotland, let alone provoked an outcry, is yet another reason why Holyrood needs George Galloway in there speaking out against the SNP's corruption of society.

* * *

Imitation is the sincerest form of flattery. The Alliance for Unity picked up a parody account from the nationalists' black ops armoury shortly after we set up our Twitter account last July. In an act of symmetry, another one popped up a few days ago for the final week of the campaign: 'AllForUnity2' purported to be a hard-line unionist account but we smelt a rat straight away when it posted a meme of British redcoats bayoneting Highlanders at Culloden. Twitter acted quickly and removed the account but it was another reminder of how untruthful politics have become in the post-modernist age.

* * *

We embark on a final soapbox tour of the South of Scotland. It is an enjoyable challenge finding something location appropriate to say in each town in a short YouTube video. To my consternation they are filmed live so there is no room for any bloomers. In Hawick I stand on the soapbox and start, 'Today we are in ....' my mind wants to say Langholm as we have just been discussing the birthplace of the Anglophobic nationalist poet Hugh MacDiarmid[64] ... but luckily my mouth says, 'Hawick'. In Lockerbie I have an opportunity to talk about the Battle of Dryfe Sands where the Johnstones butchered 700 Maxwells in 1593, a reminder of the sectarian strife that divided the South of Scotland prior to the Union of the Crowns, and which has now been rekindled by Sturgeon's divisive separatism. Outside Castle Douglas Mart I am on familiar ground and able to talk about how our livestock industries would be poleaxed by the imposition of a hard border at Gretna. After a ferocious oration by George,

---

64. MacDiarmid's real name was Christopher Murray Grieve, an early example of bogus nationalist Gaelicisation.

warning of the potential for lawlessness caused by a customs border, I point out to him that a certain section of society might not be averse to the possibilities. Stranraer provides the backdrop of the Irish Sea and the invisible but nevertheless real hard border that stubborn EU negotiators bequeathed as a memento of Brexit. I hope they have given the many thousands of viewers on George's YouTube channel a window on this forgotten corner of Scotland. But I don't think Border TV presenters need to worry about their jobs. Speaking of Border TV, we are predicted to win at least one seat in the South of Scotland but they have steadfastly refused to interview us. Why?

***

Our tour has been very revealing. Our opponents, both nationalist and unionist, have been keen to promote a narrative that dismisses All for Unity as a party that no one has ever heard of, a 'Twitter only party,' and predicts that we won't get any votes. But if the people who happen to be in high streets when we pass through are in any way representative, we are going to exceed all expectations. We were astonished by how many people say that they are voting for us. In Hawick the man from the *Times* and a photographer trail around with us. The endless stream of people wanting selfies with George is completely at odds with the pessimistic narrative – the local café owner even insisted on nabbing one of our posters to put in his window – but they write a lukewarm piece in the paper anyway, consisting mainly of Tory quotes accusing us of splitting their vote. This is not just a delusion – the diverse band of malcontents pledging us their votes has barely a Tory among them – it is a conceit. I suspect, though it may never be proved, that when the votes are counted, we will have provided a useful service by gathering up pro-UK votes from across the political spectrum from people who are disenchanted by all the major parties and would otherwise either not have voted or given their second vote to one of the many no-hoper 'none of the above' parties on the list.

***

In Loanhead we stop to pay our respects at the Miners' Memorial. The freelance photographer with us gets an exclusive of the top of George's head as he removes his fedora and bows his head for a moment's silence. I was beginning to think that not even Gayatri had seen him without his hat on. The memorial has etched on it the words 'The True Price of Coal'. There are fifteen names, an average of one fatality every two years and it is a poignant reminder of the sacrifices made below ground at Bilston Glen Colliery. George moves away and mounts his soapbox for a speech to camera on the doctrines of unity and solidarity that underpin his twin creeds of trade unionism and the union of Great Britain. We are standing near a busy T-junction and as they pass us one car in three gives us a toot on their horns and the thumbs up. It's an encouraging sign.

***

One heartening feature has been the number of people who have said that they have voted SNP regularly hitherto but won't be this time. The young mother in Stranraer spoke for many of them when she spat out the words, 'They done nothing for us. Look at it. There's nothing here for my weans.' Driving through what we assume to be fertile SNP housing schemes there is a marked absence of yellow posters in windows. Only about one person in thirty waves away our leaflets with the protestation that they are SNP voters. Could the polls be wrong? George has fifty years of campaigning under his belt, including some spectacular upsets in Glasgow, Bradford and Bethnal Green, and is feeling optimistic. If the SNP get back into power on a minority of the vote I will be very angry. The three old unionist parties have spent their time arguing over second, third and fourth places. They have barely contested the constituencies and instead focused on chasing 'peach votes' so that their candidates can slide back into their opposition sinecures on the list.

***

The campaign is challenging things on the home front. The day starts with a rush to get round the farm to see what has been happening in my absence, then into the MOT centre with my car for a re-test when it opens at 0830 – it passes, a relief, as otherwise I am stuck. There is time for a quick haircut, my first proper one since lockdown. Sheri is qualified as a physiotherapist, a stockbroker, photographer and mother (no exams for that one, just hard work and trial and error) but, though she has done a valiant job at snipping my lockdown locks, hairdressing is not part of her skill set, so it is a relief to have a short back and sides. The girls in the shop and a waiting customer are interested to hear how All for Unity is getting on, a good sign. I leave them some leaflets.

* * *

I arrive on the seafront in Troon just as George is mounting the soapbox. Getting out of my car is a blessed relief. My little car was only ever designed to be a run-around and I have been bent double for two and a half hours, all the while imagining Douglas Ross, Anas Sarwar and Nicola Sturgeon sitting on their battle buses, sipping coffee and checking their Twitter timelines. Still, there is something virtuous about being the underdog in this fight. When it's my turn on the soapbox I talk about what has been on my mind while driving. The lack of posters for any party, even the SNP, on my way here raises questions. Are the people of Scotland not going to vote? Have they lost interest in politics? Or is it that they are resigned to the outcome? Scotland seems like a woman stuck in an abusive relationship with the SNP. She can't seem to see a way out. The old parties have resigned themselves to contesting second place. Douglas Ross isn't even standing in a constituency. All for Unity has at least charted a way to defeat the SNP but are enough people listening? We'll know on Friday. Afterwards George congratulates me and says it was like listening to Lenin in Finland Square, continuing the enjoyable Right–Left banter that has become a leitmotiv through the campaign. I take it as compliment but make a mental note to compare his next speech to Oswald Mosley at the height of his powers.

* * *

Lunch at Wee Hurrie's by the harbour has to be the best fish and chips in Scotland. We are joined by a friend of David Griffiths[65] who has a Meatloaf tribute act called Pete Loaf. He shows us his act on his phone and he is excellent but like so many performing artists down on his luck after the lockdown, like Robbie, the Elvis impersonator we met in Hawick, who had had to sell his costumes to keep going. These are the people let down by the SNP's failure to get the Covid Business Relief Funds to where they are needed. I hope they get their just deserts on Thursday.

* * *

Every few minutes my phone rings with another friend asking for tactical voting advice. Usually they are from Dumfries and Galloway and the reply is simple: Lilac Tory, Orange All for Unity. I hope that if nothing else, All for Unity has helped to make people more aware of how to make their second vote work better for them. And if we do get in we will try and bring about change to the electoral system to simplify it.

* * *

Then it's off to Govanhill via Kilmarnock. Outside St Bride's Primary School the brilliant and irrepressible Niall Fraser turns up with the electronic billboard and we have a candidates' photo shoot in front of a Save Our Scotland poster. The result looks like a seventies album cover. It feels good to be campaigning on Nicola Sturgeon's doorstep, though I don't suppose she goes there very often.

* * *

---

65. A Glasgow businessman and stalwart of Glasgow Rangers Supporters' Club who is one of our candidates.

Back home there is time to catch up on social media. Linda Holt, our outstanding lead candidate in Fife has written a brilliant article in *Think Scotland* explaining why All for Unity was needed to fill the hole where the pro-UK opposition to nationalism has been for the last fourteen years. For me as a writer one of the most exciting things has been bouncing ideas off great political writers and thinkers like Linda, Effie Deans, Jill Stephenson, Mark Devlin, the editor of The Majority, whose wife Mary is one of our candidates and the political scientist Tom Gallagher, and developing doctrine to defeat separatism. That will continue whatever happens at the polls this week. I am so grateful to them, to the very generous donors who have given us campaign funds and to all our supporters who have kept us going through moments of doubt and adversity. We now have an effective campaigning force that will be ready to take on where 'Better Together' left off if need be.

* * *

Today we finish the soapbox tour outside Holyrood. George and I give our final speeches to camera for YouTube dubbed 'Krankie[66] Foes to Holyrood'. I'd love to see that headline on news-stands this weekend.

---

66. A nickname for Nicola Sturgeon is Wee Jimmy Krankie after the comic television character who bore a striking resemblance.

CHAPTER 15

Back to 'Normal'

The election itself proved to be an anti-climax, as it had been increasingly apparent it would be as the campaign progressed. The charge of vote splitting by the major parties stuck. The lack of mainstream media coverage made it very hard for us to put our message across and lack of funds meant that we didn't manage to stuff enough leaflets through letterboxes. Covid restrictions meant that we were unable to hold the type of rallies at which George excels. There are significant barriers to entry for small parties as the media can turn round and say that they have not demonstrated sufficient support to receive coverage and the only way to generate sufficient support is through the media.

Nevertheless we received 0.9 per cent of the vote and came seventh just behind Alex Salmond's Alba Party. It wasn't enough to win a seat. We were well ahead of the other small parties such as UKIP and Richard Tice's Reform Party. The result did partially vindicate our decision to try to break the stalemate – by voters sticking with their party loyalties Scotland ended up with almost exactly the same result as last time, another five years of SNP domination, despite only 31 per cent of Scottish adults having actually voted for them. The only difference was that the SNP had to go into coalition with the Scottish Greens to form a government, which is even worse news for the countryside and for anyone worried about the Green's woke agenda. The policy of encouraging primary school children to reconsider their genders and a recent survey of all fourteen-year-old schoolchildren in Scotland asking whether they had anal sex regularly gives a chilling insight into the way things are going.

We had hoped that our intervention would make the larger pro-UK parties work harder to win constituency seats but unfortunately it had the opposite effect. They ran very negative campaigns with 'both votes' slogans to try to get their MSPs back into the fifty-six list seats, virtually guaranteeing that the SNP would win nearly all of the seventy-three constituency seats and therefore the election. Both the Conservatives and Labour had fobbed us off with assurances that they would deploy nudges and winks behind the scenes to ensure that they didn't campaign too hard in each other's safe seats. In the event, the Tories put a huge effort into East Lothian and Labour campaigned extra hard in Ayr with the result that both seats fell to the nationalists.

In the final analysis it was a good cause, we did our best but it wasn't enough.

However, I have been told by a number of people that we did succeed in changing the overall conversation about independence and that the resistance to separatism is now much more robust as a result.

We all reverted to our previous lives, I, with some relief, back to the obscurity of farming in deepest Galloway, George and his Workers Party to woo the electors of Batley and Spen in a by-election with his unique blend of wit, fire and brimstone.

Someone else will have to come up with a better idea to break the stalemate in Scottish politics next time.

* * *

'Did you remember to wet the barrel?' The exigencies of parenthood can be summarised by a series of questions from 'Have you changed his nappy?', to 'Did you remember to feed the guinea pigs?', to 'Have you paid the school fees?', to' 'You did lock the drinks cupboard didn't you?' to this. We were initiated into the forgotten art of cooperage by our son Oliver, now, since lockdown, an entrepreneur with a start-up rum business. 'Do you mind taking delivery of an empty malt whisky cask and looking after it?' sounded a sinecure until we discovered that empty barrels have to be kept damp to stop the wood drying out and shrinking. It's amazing how having children has expanded our skill sets. It has also turned us into seaweed gatherers. We had long wondered how we could monetise the luxuriant fronds of spiral wrack and dulse growing on our rocky shoreline. I like eating it in the spring as an alternative to green vegetables but, as I couldn't convert the rest of the family, the market seemed limited. I had high hopes from the growing demand from new technologies to save the planet by reducing cow flatulence, but I'm told it's the wrong sort of seaweed. Besides, without the possibility of mechanical harvesters, it has to be hand gathered, raising visions of those Victorian oil paintings, by the Glasgow Boys, of kelp gatherers with hand carts, that romanticised back-aching work by impoverished crofters after the Highland Clearances, so the margins seemed tight. Then Oliver created *John Paul Jones Lowland Rum* after discovering that Caribbean rum tastes delicious after being steeped in Scottish seaweed, inter alia (I can't divulge the exact recipe), and even better aged in old whisky barrels for the premium product. Bingo. The original John Paul Jones, rebel pirate or heroic founder of the American Navy according to viewpoint, was born on this estate in 1747 and it may have been a childhood spent gathering seaweed that persuaded him to cross the pond to seek his fortune. It is heartening to think that Oliver may have a valuable business to bring home with him one day, and in the meantime he is gaining valuable business experience that will stand him in good stead when he has to take over from me.

The Chief Photographer is now to be seen snapping bottles and barrels bobbing around in the surf like extras in a *Poldark* episode for the PR department (daughter Rosie), when she is not picking seaweed or wetting casks. The last time spirit barrels were seen washing up on our shore may have been during the smuggling boom in the 1700s when the Galloway coastline was the hot destination for duty free booze from the Isle of Man. A contemporary Customs and Excise report on a haul of brandy seized right in front of our house states, 'We could not say whether the laird was involved but many of his servants and horses were.'

The long drought has come to an end now, meaning that the barrel can be kept moist by the Almighty until 200 litres of golden nectar arrive to wet it from the inside. The natural world seemed to put spring on hold until the rains came but now there has been a marked increase in birdsong as the leaves slowly unfurl in blustery hailstorms. The swallows kept us in suspense right up until the end of April before venturing this far north. Spring here is always partly characterised by finding ducks in strange places. Mallard are seen creeping along ditches again and splashes in fields are graced by elegant pairs of shelduck. It has officially been the coldest spring of my lifetime, after a run of them in recent years. We had almost forgotten what frosts looked like down here on the coast but they have been a regular curse this year right into May. It does seem that there may be an inconvenient truth in the mutterings from scientists that the planet is entering a Grand Solar Minimum, or mini-ice age, that could dominate the next few decades. This is contrary to the prevailing wisdom of the broadcasting media so it is hard to find out much about it. Will I soon be taxed for daring to have cattle while my son may be paid to have more of them to push life-preserving carbon dioxide into the atmosphere? Or are these the rum-soaked ravings of a heretic? Who knows?

***

'Oak before ash in for a splash; ash before oak in for a soak.' The oaks have definitely won this year, and are delighting us with the bright, pale-orange complexion of their early foliage before it rapidly takes on its summer green, a foretaste of their flaming autumn colours. The ashes are a different matter altogether. In fact, I fear we may be waiting in vain for some of them to take leaf. *Hymenoscyphus fraxineus* or chalara ash dieback has reached us. Some ashes are struggling to come into leaf; patches of green are showing on wintry skeletons and there won't be enough to nourish themselves. Death seems inevitable. John echoed my thoughts when he said, 'Why couldn't it have been in the sycamores?' The place won't be the same without ashes. They stud our hedges and I regard many of them as old friends. I love watching them ripple like the sea on windy days. Sycamores on the other hand grow like a weed here and we have to waste valuable time thinning them to stop them taking over and shading everything else out on the woodland floor beneath. I like them well enough in maturity, and we have some fine old ones here, but they don't have the elegance of the ashes and the firewood is not nearly so good.

We have known that it was coming for some time. Like other pandemics: Dutch elm disease, squirrel pox, *Phytophthora ramorum* in larches, it has rampaged across much of the country before hitting us here on our remote peninsula in Galloway. For several years we have reversed our usual thinning priorities and favoured other trees over ashes – even birches and sycamores. Now that it is actually here we can only hope that some of them might have some genetic resistance and survive to keep the line going. We are cherishing them while we still have them as one might look upon an elderly relative with a terminal disease. And there are difficult decisions to be made. Two whole woods my father planted with predominantly ashes may have to be felled. The trees are at the age where they seem most susceptible, around twenty-five years old, and their bark has ugly black blotches on. And the evidence is that, as might be expected, where the planting is densest, so also is the disease most concentrated. Single older trees seem more resilient. On the other hand, old trees with spreading limbs may be harder to fell once the disease takes hold. The advice is that the wood may become brittle so that it is no longer safe for tree surgeons to climb. My inclination is usually to let nature take its course. I have often seen trees recover from various ailments, and standing deadwood is fantastic habitat in any case. But there are too many of them to risk it and the disease apparently ruins the timber.

At the risk of sounding like Donald Trump, the fungus that causes ash dieback is a Chinese disease. It doesn't kill the two species prevalent in China, Chinese ash and Manchurian ash, which manage to cope with it, but has a devastating effect on our native British trees. It appears to be the arboreal equivalent of squirrel pox. The answer eventually will be to restock our woodlands with one of the disease resistant species but so far these are hard to source for the good reason that no one wants to compound the error by importing trees carrying disease. We have had some success with disease resistant elms, *Ulmus Columella*. But it looks as if for the rest of my lifetime we will be dealing with dead and dying ash trees.

June 2021

My eye is caught by a large flying object and for a moment I fear that a glider plane has flown off course and is about to crash into the castle. Then my brain registers the stork as it deftly weaves its way around John Nash's crenellations to land on the nest and relieve its mate of looking after the chicks on a faggot of sticks that would have tested the strength of Good King Wenceslas's yonder peasant. Other country house owners have jackdaws in their chimneys, über conservationist Charlie Burrell and his wife, the writer Isabella Tree, have *Ciconia ciconia*, the European white stork, blocking a flue at Knepp Castle, their home on the Sussex Weald. Knepp is at the cutting edge of the agricultural counter-revolution and for Charlie and Issy, and increasing numbers of landowners of this generation, the challenge is to be able to show storks where none bred before, at least not since the time of Henry V (perhaps the 'happy few' scoffed the last one at a post-Agincourt banquet).

I am there partly to research an essay I am writing for the Charles Douglas-Home Memorial Trust, who have generously given me an award to write about 'wilding'. And partly driven by a raging Noah Complex. This is an entirely self-diagnosed condition I made up on the spur of the moment recently when, as so often happens in a debate, it popped up in the back of my mind as a way to communicate the way that farmers feel about wildlife to my fellow festival panellists, particularly a vegan professor of sociology and a philosopher-poet who felt the need to give us her personal pronouns (she/her) – I suppose you can't be too sure these days. Noah Complex is particularly prevalent among farmers over fifty and consists of a nagging fear that the species present when we started farming will not be there at the final reckoning and we need to have more on our own particular ark. Visiting Knepp is a Damascene experience. At home I flatter myself that the birdsong is like a symphony played by an orchestra, when, in fact, I am often probably only listening to a quartet and only where there is good cover. At Knepp, even in the middle of the day in a light drizzle, over the whole estate it is like listening to the Massed Bands, Pipes and Drums of the Household Division playing one of John Philip Sousa's most exuberant marches. It is loud, the sound of synergies being exploited through innumerable symbiotic relationships from deep in the soil to the tips of the large oak trees, several of which also contain stork nests. It is a challenge to do better but also reassuring that Mother Nature can restore the ecosystem to a healthy balance after twenty years of minimalist livestock farming that has allowed the scrub to take over to the point where on a hot day the swallows must sometimes wonder whether they are still in the African bush.

Back home, I head straight to check on my own conservation efforts. My Noah Complex has been most acute over ground-nesting birds, particularly waders. It is easy enough to provide habitat for tits and thrushes in the woods and for warblers and yellowhammers in the hedges, but protecting birds from cows and crop cultivations in fields is the hard bit. No doubt soon technology will allow us to plot nests electronically so that satellites can guide machinery and grazing animals away from them, but we are not there yet. Last year I took the decision to fell two belts of trees that were allowing crows' perches to dominate what would otherwise be perfect habitat between the burn and the sea – wet in places and a long way from the nearest badger sett. We were already also trapping the crows hard and this spring it has worked. I can hear a cuckoo calling and spot an oyster catcher on her nest, shelduck, ringed plover, a pair of greylag geese, larks, hares and best of all, there, in the middle, three small, perfectly formed lapwings flanked by their parents, who rise volubly every now and again to mob any rooks that venture too close. I feel as Noah must have done when the dove brought him the olive branch as a sign of hope.

* * *

I am suffering from the chronic seasonal condition known as game crop neurosis (GCN). With the dry, cold conditions this spring it has been particularly acute this year. The game crops form a tiny percentage of the agricultural activity here but somehow manage to fill whichever part of the brain processes anxiety, elbowing out of the way the potatoes, cereals and grass on which we rely for our daily bread. It's not just that they keep pheasants where we want them to be, in our 'signature drives'. Their success or failure will be visible to a far wider constituency than the other arable crops. If they fail, there is the consolation that they will only hang around for a few months to taunt us. The game crops will look dreadful all winter provoking sniffiness in the beating line: 'Och, it's hardly worth doing that drive this year, there'll be nothing in it.' And hoots of derision from my fellow guns at elevenses: 'Bit of a cock-up on the kale front this year was there?'

It's awkward to admit it, as it plays into the hands of our enemies, but we are facing a biodiversity crisis. The more successful we are at growing uniform crops in our fields, with barely a seed-bearing thistle in sight, the fewer opportunities there are for nature and therefore the more I want the wild bits to froth with nettles and teem with every native species in my well-thumbed copy of Collins *Complete British Wildlife*. I have started to focus on the farm within the farm – the woods, hedges, walls, ponds, bogs, streams, reed beds and patches of thorny scrub and wild-flower meadow. And the game crops are a critical part of it as many of the birds that breed in the aforementioned bits rely on them for their winter food. I have been chastened and enthused by my trip to see Charlie Burrell at Knepp, and another to see the writer James Rebanks on his Cumbrian farm where he is successfully integrating a profitable beef and sheep enterprise with wildlife. His concepts of treating every field as a woodland clearing and never being more than 300 feet from another habitat are starting to have visible effects.

Back home the GCN is compounded by the need to get the crops that make the money in the ground before the game crops, particularly as my priorities are exactly reversed in my share farming partners. This isn't always a bad thing. The temptation to go too early can leave tender seedlings at the mercy of late frosts – as happened this year on some farms, as one arable contractor told me with what I thought I detected was a degree of satisfaction – but leaving it too late can be disastrous in a dry summer. This year we stuck to the old adage and didn't cast a clout till the may was out and sowed in the last week in May, then waited anxiously through June as first nothing happened, then slowly plants emerged and were sustained by the dew and occasional shower before the Almighty took pity on us and sent a good downpour on 25 June.

Now there is a satisfying green cover and I can see kale, quinoa, red clover, oats, chicory, phacelia, mustard and vetches mixed with wild redshank, fat hen, ox-eye daisies, coltsfoot and charlock. Most of it may be artificial wilding but the birds won't be fussy.

July 2021

I wake disorientated. Outside I can hear barnacle geese barking and the high pitched peeping of oyster catchers and my inner satnav is telling me that I am at home on the Solway Firth in winter, but the vastness of the four-poster bed and the old master of some biblical worthy on the wall and the warmth of the July sun coming through a chink in the thick tapestry curtains are telling me something completely different – and then I piece together a night of watching Eng-er-land beat Denmark, amply lubricated by copious quantities of Pomerol's finest, and I pull back the curtains to see house martins dancing across the Palladian façade of one of England's best-loved stately homes and gaze across the park at fallow deer tiptoeing from the trees to drink in a lake covered in water fowl. It is a vision of Eden. I remember my host telling me that the barnacle geese had arrived in 2009 and now breed in the park each summer rather than taking the long trek back to the Arctic Circle each spring. Why wouldn't they? They winter elsewhere, possibly on the Solway, as they would soon discover by tagging them.

I'm at Holkham; as part of my picaresque quest to understand how the British countryside is managed, I have made the pilgrimage to Norfolk to the wellspring of the agricultural revolution, where Thomas Coke, 1st Earl of Leicester, innovated crop rotations. His descendant and namesake, Tom Leicester, a school and army contemporary of mine, is also passionate about the farmed environment and leads an energetic team imbued with his determination to cherish the estate's natural heritage and reduce reliance on chemicals by extending rotations. Where once farm tours at Holkham would have focused on crop yields, now farmers like me go there to see how highly profitable food production and successful wildlife conservation can go hand in hand. Thick crops of barley in the middle of the fields are mere wallpaper as we obsess over the extra-wide hedges ('hedge porn' in conservationist speak) and hay meadows on margins brimming with fumitory, favourite snack of the turtle dove. Tom's ebullient Director of Conservation Jake Fiennes guides me around the Holkham National Nature Reserve, the same disconcertingly blue eyes as his more familiar thespian siblings flashing with enthusiasm for his subject. Once managed by Natural England, the reserve has now been taken back in hand by the estate and is thriving under Jake's robustly undogmatic husbandry that blends scientific rigour with pragmatic farming, forestry and predator control. For me this is the ultimate busman's holiday, and it is a joy to see how carefully managed grazing by cattle and sheep is producing healthy populations of lapwings, avocets, marsh harriers, natterjack toads and many more threatened species. It is a reminder that for all the much-trumpeted busyness of the quangocracy and certain powerful charities on the land, the agricultural counter-revolution is being driven by practical landowners and farmers on the ground and not by environmentalists in offices and television studios, and is often happening despite them rather than because of them.

Ignoring all the noise from prominent Neo-Roos in the media, wilding has thus far been landowner led. This has been a tweed revolution not an anorak one and it has been bottom up not top down. High-profile BBC presenters and *Guardian* columnists, civil servants, quangocrats and charities have much of the power but very little of the day to day responsibility for the land – that rests very firmly with farmers and landowners, a group that curiously has been marginalised in the debate and only finds itself occasionally represented by token 'practitioners' on the interminable committees and steering groups that increasingly seek to shape the countryside. If anything, wilding has shown up the fallibility of the charities sector and the quangocracy. The Holkham NNR achieves much better results for its wildlife than it did under Natural England, partly because Jake takes a more robust approach to farming the marshes and predator control. Similarly on Elmley Marshes in Kent, where one half was managed by the RSPB and the other by the farmer Philip Merrick. The success of breeding lapwings was measured in a scientific study. The RSPB achieved 0.11 fledged chicks per breeding pair, well below species sustainability. The Merricks' land achieved 1.1, a level at which the species can increase. It seems that the environmentalists do not always know best.

***

I stayed one night in Norfolk with an old army friend David de Stacpoole and his wife Jane, the daughter of Michael Bratby, whose wildfowling and punt gunning exploits on the Wash with his friend Peter Scott are immortalised in the books they wrote together. David and Jane have carried on the ornithological tradition and David took me to see where he shoots on the marshes where game conservation and the rearing of rare birds like turtle doves, stilts and bar-tailed godwits for release into the wild go hand in hand. An idea that other shoots could perhaps consider. After dinner they showed me their party trick. Some meat was placed on the bird bath and we waited in the dark for a few minutes and then a tawny owl swooped from a nearby tree, deftly picked the food up in its talons and flew off with it. Twelve years before, the de Stacpooles had rescued a fledgling owl and brought it up before successfully releasing it back into the wild. They had carried on feeding it and a few years later the original owl had appeared at their door stricken with poison seeking their help before dying at the vet's surgery. But its children and grandchildren had continued to come to the food – no doubt successfully protecting the local partridge population each summer in the process by 'barrier feeding'! It was a memorable wildlife experience that put the seal on a wonderful trip.

***

As soon as I am home, as farmers are wont to do I drive round to see how my own attempts are matching up. Reassuringly, the hedge backs are frothing with meadowsweet and violet clumps of vetch, and a brood of linnets is perching on a wire beside a cow track. First stop is the newly seeded grass where until this spring there had been thick rushes. The history of agriculture can be deciphered in the soil strata of Catie's Field. It had always been good dairy pasture until the 1970s; then the EU's Common Agricultural Policy drove it on a journey through chemical-soaked wheat to ragwort flecked permanent set-aside that became ever thicker rushes, when it seemed easier to be paid to grow nothing in the unforgiving clay. My father said that he knew he had been beaten by the harvest of '85 when he saw a wisp of snipe rise in front of the combine just as he got it bogged. You can still see the outline of the scars it left. Eventually I brought beef cattle on and wrestled with the rushes and battled simultaneously with the civil servants to tick boxes in an environmental scheme that has now run out, seemingly without any replacement. Finally we decided this dry spring to flail, bale, spray, rotovate, shallow plough, subsoil, lime, power harrow, roll, drill, fertilise and roll it again back into how it was in 1973. The new style of regenerative dairying with rotational grazing that maintains insect-rich soils, and a mosaic of grass, plantain and clover leys at varying heights, may be more likely to support wildlife anyhow. A hare is picking her way delicately across one corner. It seems a good omen.[67]

***

The shudder across the countryside is audible as the government trumpets the announcement of a free trade deal with Australia. Diehard remainers are tweeting, 'I told you so.' And there will be long faces at the mart. As I am a farmer who voted to remain in the EU you might expect me to be 'Disgusted of Dumfriesshire' but, somewhat to my surprise, I am sanguine about it. Brexit is a done deal and we have to make the best of it now. Tilting at windmills is fun for columnists but health-threatening for farmers. If Sir Robert Peel was able to face down the landed interest in 1846 to repeal the Corn Laws for the greater good of cheap food for the consumer, it is unlikely that a government elected on a manifesto to 'get Brexit done' will put the interests of Tory farmers before those of hungry voters in the 'red wall'.

---

67. It was a good omen. At the time of editing it is covered in curlews and lapwings.

There is also part of me that thrills at the achievement and the opportunities for 'Global Britain'. This is by any measure a stunning achievement by Liz Truss, and the bigger prize of entry into the wider Asia–Pacific free trade agreement now seems that much more achievable.

There is still a large section of the population that doesn't get Brexit, largely because they see it through the prism of a still Europhile media establishment. This is having particularly toxic effects in the devolved nations where Brexit is still used as a jemmy to prise the union apart, and even the Scottish Tories can't bring themselves to embrace it. Most Britons did not listen to the Prime Minister's seminal Greenwich Speech in February 2020 because the BBC and other media barely reported it. It is worth finding on the internet:

> We are embarked now on a great voyage, a project that no one thought in the international community that this country would have the guts to undertake, but if we are brave and if we truly commit to the logic of our mission – open, outward-looking – generous, welcoming, engaged with the world championing global free trade now when global free trade needs a global champion, I believe we can make a huge success of this venture, for Britain, for our European friends, and for the world.

The trade deal needs to be seen in the context of that vision. In the same speech Boris bemoaned the fact that though we can now export beef to China, 'still no lamb, not a joint, not a chop, not a deep-frozen moussaka, even though we have the best lamb in the world.' He then went on to point out that Wales is closer to Beijing than New Zealand, a bold assertion that had everyone scurrying for their atlases. After forty-seven years of reliance on the EU internal market and the Common Agricultural Policy, it takes a leap of faith to recognise that there is a vast global population out there of largely carnivorous and increasingly affluent consumers who cannot grow beef, lamb or dairy products in their own countries, many of them members of the Commonwealth. The potential upside from Brexit has yet to be fully examined let alone exploited.

But that doesn't mean that the Australian trade deal is not fraught with danger for certain sectors. It is a huge relief to be out of beef, although ironically we are now producing more meat out of the dairy then we were before.

* * *

As viewers of *Clarkson's Farm* will have gathered, farmers don't watch much television, certainly not during the summer, so when the *Daily Telegraph* want a farmer's eye view of the programme it involves an immediate 'binge' session on Amazon Prime and the hurried assembling of various rustic 'focus groups' to discuss the series. The response is overwhelmingly favourable. A high percentage of farmers think that Clarkson had done farming a great service, a figure only exceeded by the number of male farmers expressing a desire to make the close personal acquaintance of Clarkson's pulchritudinous blonde partner, Lisa from Dublin. Yes, we can carp about the realism of being able to employ a shepherd to look after only seventy ewes (or 'lady sheeps' in Clarkson's argot) and how new buildings and machinery were financed on the back of Jezza's TV earnings. Farmers in North Dumfriesshire (annual rainfall 90 inches) are also scornful of Clarkson's attempts to portray weather events on his easily farmed 'boy's land' in North Oxfordshire (annual rainfall 29.7 inches) as apocalyptic. But what is uncomfortably close to the truth is that a high proportion of farms we know contain a hapless public-school-educated halfwit who pays all the bills, routinely gets trodden on or kicked in the gonads by livestock, and puts up with a stream of invective from all the other team members, especially his employees, when he breaks machinery or generally gets in the way, and only earns around forty pence per day, if he is lucky, for a hundred-hour week. And that this bittersweet masochistic existence is driven by a misty eyed romanticism about his farm, its flora and fauna, and the way of life it allows. This observation is particularly pronounced among colleagues on my own farm, many of whom comment, 'Jeremy is just like you.' This statement of correlation is most acute when suffixed by the word 'Dad'.

* * *

'Wilding has moved the conservation arguments on a bit.' I am not in the habit of quoting RSPB managers. Their idea of conservation is very different from mine. But the more I delve into the issue of wilding, and look at different approaches from Sutherland to Sussex, the more I have to admit that there is some truth in that statement (by the manager of RSPB Loch Garten, of osprey fame). My attempts at creating a paradise for game on my small patch are helped by the latest ideas from places like Knepp Castle about the importance of biodiversity and a balanced ecosystem built upon healthy soils. Conversely, there is no doubt that the wilding of Britain, which has been going on all along, though you would not get a *Springwatch* presenter to admit that, has always and continues to be helped by the desire of landowners who shoot to add a bit of variety to the bag. It was partly with that in mind that I decided that this summer, while the bank manager is suitably reassured by the price of milk and the queues of tourists wanting to book our holiday cottages, I would launch Operation Dredge.

We had five ponds notionally, but one had gummed up completely, the result of being dug too shallow last time, and the grass growing up through the water then dying back each year, so that it was really no more than a rushy patch. And the other four were going in the same direction. One was designed as a silt trap, a dogleg on a big ditch that caught our topsoil before it washed into the burn and then out to sea. It had been doing its job rather too well. Two had so many bulrushes that you could no longer see any evidence of water. Another had disappeared under willow scrub. They were a reminder that nothing stays still in nature for long and that, without men with diggers, there are few natural causes of ponds and even fewer for preventing them being choked by the Triffids and ceasing to be ponds. These are inconvenient truths for the wilder fringes of the wilding movement, who view any human intervention as wrong.

But as of this week any passing duck will have its eye caught by ten gleaming expanses of water, all with islands for safe roosting. We had various boggy bits in woods where the soggy roots manifest themselves above ground in sickly looking trees that are dying from the top down. It is nature's way of telling us that we are trying to do the wrong thing there and might as well try something different. A harvest of the occasional teal might be better than a barren harvest of timber that is hard to extract through the mud. So the trees came out and the ponds went in. And with them I hope damsel flies, dragonflies, frogs, toads, newts, herons, moorhen and so on, and with luck mallard breeding in the spring and teal feeding in the winter. We will be doubling the aquatic habitat here and it will be interesting to see what effect it has on certain species. As we completed the first one, the air above filled with swallows.

* * *

'I'm afraid we'll no be here for a few days.' Willie looks a bit sheepish as he goes on to explain, 'We had a bit of an accident yesterday.' My mind races around various possibilities. Of all the operations that happen on the estate, with the possible exception of loading cattle onto trucks, felling trees is the most dangerous. Foresters are a hardy breed and all too many of the ones I have known have carried some disfigurement with them as badges of their trade, a permanent reminder of the need for health and safety in the workplace, even if, or rather, particularly if, your workplace happens to be sixty feet up a tall beech tree. Jock, who tended our woods when I was a child, was missing a little finger. I used to stare at it, with hindsight probably rather rudely, while being given the local delicacy, home-made tablet,[68] in their cottage by his wife Nancy, who was our nanny. He had cut a piece of timber rather too fine and pushed it through the circular blade in the sawmill. Family legend has it that my grandmother packed the severed digit in ice and drove him to the hospital but they were unable to sew it back on. Then there was Jimmy, the first tree surgeon I ever employed, who had one glass eye. Early in his career a tree landed on his head; his helmet had saved his life but not his eye. Then Ainslie, our other regular forestry contractor, gave everyone a fright on a neighbouring estate recently when he stepped back with his chainsaw still running, tripped and fell flat on his back. The saw landed on top of him and sliced through his helmet visor taking out one eye.

---

68. Pronounced 'tarblet,' an extra-sweet form of fudge.

Fortunately, this accident was minor by comparison. Daniel, the other half of the team, had been helping Willie and, with the over-confidence that comes with familiarity, had put his hand in to pull away the sawn branch a fraction too early and caught Willie's chainsaw with his little finger. The protective glove had done its job and stopped the chain going round in the nick of time so that it bounced off his knuckle, otherwise he might have lost his hand. Its bloodied remains were still on the dashboard of the pickup. It was a lucky escape; A & E in Dumfries was able to save his finger with fifteen stitches – but a salutary reminder that chainsaws are lethal, and even those who use them every day can fall prey to them.

This year's thinning operations have given me more satisfaction than almost anything else. People say that they love planting trees but in my experience the act of planting itself, as opposed to the planning and dreaming that goes with it, is rather a dull chore for which I have little patience. The real excitement comes when a new wood gets its first proper thinning and you can see both the wood and the trees for the first time. This year three woods my father planted twenty-five years ago have had a major restructuring. The hardwoods were planted, as was the custom until recently, with a nurse crop of conifers, mainly Norway spruce, in alternate rows. The spruce made the hardwoods grow tall and straight towards the light but the woods themselves looked and felt like dark conifer plantations, until now, when most of the spruce have been removed and we can see the oaks properly for the first time and get a sense of what the woods will be like when they are mature, with the dappled forest floor carpeted with wild flowers. It is a sadness that my father is no longer alive to see them.

We have not planted any more conifers for years now, with the exception of Scots pine. In fact most of our woods are now regenerated naturally with the oaks mostly being planted by jays I suspect. I have applied to the woods William Morris's dictum: 'Have nothing in your house that you do not know to be useful, or believe to be beautiful' and culled the less useful (for wildlife) or less beautiful trees like sycamores mercilessly and planted others that were missing. My descendants may not thank me for reducing the timber yield but I hope they will enjoy a landscape brimming with wildlife.

Meanwhile down in the timber yard next to the dairy there is a pleasingly large stack of ten-foot logs. It will serve as a windbreak through the winter then nesting habitat for birds through the spring before the firewood processor turns them into fuel for our biomass boiler next winter. Forestry is a long game.

* * *

'Don't step out of the back of your butt to pick up or you will disappear. Enter and exit from the side!' My host's advice might have included the words 'And don't forget to put some midge repellent on.' Fortunately, Mark, my long-suffering loader for the day, had some, although, despite applying it liberally, we were both itching while we waited for the first covey to appear. My eye kept being caught by a wide variety of flying insects, which for a nanosecond I took to be grouse each time, including one that looked like a mini-microlight. I reflected that if I were granted a wish right then I would opt to make certain high-profile media personalities sit there for a couple of hours, minus guns and midge repellent. Then they could make a close inspection of the blocked 'grips' and the wet patches of bright green moss atop bottomless peat that would swallow a cow whole, which hold the water in a giant sponge then release it slowly, filtering it so that it comes out of a tap in Huddersfield with perfect quality; or at least it would if they didn't first treat it with chlorine. And they could marvel at the rich smorgasbord of insect life on a wet grouse moor and understand how it supports all those lapwing, curlew and golden plover chicks. So that they could then reflect on the deliberate untruths they spout about grouse moor drainage causing floods in Yorkshire's towns and cities and atone for them by itching dreadfully.

Looking across a patchwork of bird-rich habitat managed by decades of burning, one can appreciate the care that goes into maintaining that part of the Peak District for wildlife. The keepers and loaders are worried about the risk of wildfires now that it is no longer permitted to burn fire breaks, let alone the small 'cool burn' patches needed to give grouse territories a variety of habitats. They are all veterans of firefighting incidents when they have stood shoulder to shoulder with the fire brigade fighting fires on moorland that has not been managed to reduce the fuel load. Technology has come partially to the rescue with a machine that mows and mulches but the re-growth didn't look as good as on burnt patches, and there is still flammable material left.

It was great to be in action again, a little slower up the hill and my reactions perhaps not as sharp as they were, in fact definitely not as sharp as they were, but so grateful for the opportunity and the chance to meet up with friends I have known for over forty years. All went blissfully until the last drive when a covey crossed in front. I fired, missed and swung faster, confident of bringing the lead bird down with my second barrel but nothing happened. Not even a click when I pulled the trigger. The next time I changed back to that gun the same thing happened again. With growing dismay I realised it was probably the spring rather than a firing pin.

Fortunately Barry the wizard of Upper Nithsdale can fix it. Without his skill as a gunsmith my grandfather's old guns would have packed up years ago. I am going to have it mended in the hope that the cartridge manufacturers will come up with a viable solution to the ban on using lead, when it comes. I have yet to see any science that proves that the ban is really necessary. However, the countryside unions seem to have rolled over on the issue without a fight. The modern over-and-unders appear to be more robust with fewer things that can go wrong, so I fear gunsmiths like Barry will have less work. But dentists will no doubt prosper from the repairs to broken teeth after they have bitten on steel shot.

* * *

I haven't checked with Cumbrian friends but I suspect the explosion may have been audible across the Solway (given that we hear music from Aspatria Festival each summer; it's amazing how sound can travel over water). The Mistress of the Household was just packing for our family holiday in Mallorca, the first since the plague, when our cleaning contractors broke the news. One of them had been 'pinged' by the Covid app. Sorry, but we would have to make other arrangements. Visions of missing our flight so that we could muck out holiday cottages tormented us.

We hastily rang round and assembled an ad hoc cleaning team and the holiday was saved. It is at times like these that the word 'community' has special meaning.

* * *

I am told that foresters in the south of England are now worried about the European spruce bark beetle (*Ips typographus*), a type of spruce-destroying weevil. Fishing on the Spey in the last week of September I have to confess to wickedly thinking that a bit of destruction might not be such a bad thing in the blankets of Sitka spruce covering the upper catchment. It has been a frustrating year for fishermen in Scotland with drought conditions making it a poor season on many rivers. Then, at last, the rain came on our first day, but in such quantities that it soaked through my cap and the river rose two foot and turned mocha coloured. And the fish, with silt in their eyes and gills, not unreasonably went on strike. Over breakfast on the second day, before hitting the first tee at 'Royal' Spey Bay, Angus our host explained how in our youth, forty years ago, the river would have risen more slowly and with less sediment. Then the forestry ploughs turned the heathery sponges in the hills of Upper Speyside to dirty corrugated roofs. All that stuff we learnt in geography lessons at school about trees slowing down the flow of water was demonstrably piffle. The water in our rivers has also been acidified by the conifers, one of many probable causes of the steep decline in salmon numbers to the point when many are openly talking about their extinction in our lifetimes.

Researching my wilding essay during the summer I saw, on Derek Gow's West Country farm, how beavers can have miraculous effects on the upper reaches of rivers by braiding them and creating dams that filter water, and small channels that benefit spawning fish and insects that feed their young. Less positive are the beaver impacts downstream and I can imagine them blocking drains and breaching flood banks that protect the barley needed for the distilling of Speyside malts, as they have done on the Tay. Seeing the beautiful beech avenue in the park at Gordon Castle, planted as a memorial to the young men of the estate who never returned from Flanders, I reflected that it would not have reached its centenary if there had been beavers around. It's a dilemma I suspect many riparian owners are going to have to confront before too long.

The disappointing end to the season was a shame for the Head Ghillie, Ian Tennant, who was retiring after forty-one years at Gordon Castle and the Brae Water.[69] The embodiment of all that is good in the countryside, his passion for the river, infectious enthusiasm and knowledge meant that fishers have loved being with him and many returned year after year as a result, to the benefit of hotels, shops, petrol stations and the wider local economy. He was singled out by the Duke of Rothesay, as we know Prince Charles in Scotland, as one of his six 'countryside heroes' when the prince guest edited *Country Life*. In his retirement Ian is going to be a 'river bagger' (akin to a Munro bagger). He wants to fish every river in Scotland for charity (The Atlantic Salmon Trust and Prostate Cancer UK).

Then, on our last day, the last of the season, a fisher well known to Ian over the years came to fish and with Ian's help landed – and put back – an eighteen-pounder. The smile on Ian's face at lunch was as broad as the Spey itself. Proof, if one ever doubted it, that the Almighty is a fisherman, and a royalist.

* * *

It's a mixed-species metaphor but the sky is black with chickens coming home to roost. There is a pig slaughtering crisis. Not for Remoaner reasons but because the Brexit government has so far failed to take its own advice about a bonfire of regulations. There are pigs backing up on farms for lack of abattoir capacity because they can't get the staff[70]. Many of us have said for years that our food system is made vulnerable by increasingly relying on fewer, bigger abattoirs caused by EU regulations that favoured the Big Food corporates and drove the smaller slaughterhouses out of

---

69. During which over 240,000 rod days have been fished there and nearly 60,000 salmon caught.

70. And that is partly because of Brexit. The removal of Eastern Europeans from the labour market has not so far been matched by British workers coming off benefits and into work in manual jobs like slaughtering.

business. Two generations ago every large village had a butcher's shop with a small abattoir behind it where local livestock would be killed then butchered into cuts to be sold over the counter wrapped in brown paper. During our time in the EU most went bust, to be replaced by gigantic factories that processed meat into plastic-wrapped products for supermarket chillers. The costs of gold-plated building regulations and complying with paperwork to appease a growing army of government agencies, even the requirement to have vets present, proved too much. Small farmers were also affected; the lack of competition among buyers has steadily eroded the farmer share of the retail price.

There needs to be a quick fix to put bacon back onto British breakfast tables. Derogations are required urgently to allow farmers to slaughter pigs on farm so that pig meat can go through the food chain, via butcher's shops and temporary butcheries, rather than – disgracefully – into landfill. It is a mystery to me why I can legally go out on my farm and shoot a roe deer or a pheasant and cut it up for the pot but not a bullock, a pig or a lamb. We need to stop thinking that civil servants know best and re-empower farmers to do what they always did before we allowed the EU to tie us up in knots.

Longer term, there needs to be a thorough overhaul of our food system. When the regulations covering the distilling industry were reformed, a thousand micro-distilleries bloomed. The same thing needs to happen in the meat industry. The lobbying, the threats and the dire warnings of animal cruelty and botulism from vested interests in the food cartels and the civil service will make this hard for government but they need to act or deal with a shortage of sausages every time there is a hitch in the supply chain.

Surely this is just the type of problem that Brexit was supposed to solve?

October 2021

I was driving round looking for signs of teal yesterday when I came across a middle-aged woman rummaging in the undergrowth in one of our woods. It was close to where I had recently picked up a fly-tipper's pile of rubbish so I accosted her. 'It's okay, I'm geocaching' was the cheery response. Having led a rather sheltered life I needed further explanation. It turns out that there is a fanatical sect of über-ramblers that goes around hiding small containers and logging the coordinates so that people with nothing better to do can use the GPS on their phones to find them. Those doing the hiding are supposed to ask the landowner's permission, she told me, as she showed me a cluster of icons on a map on her screen. It seems a harmless activity but, as I explained to her, would be very damaging to wildlife during the nesting season. Perhaps if some official in one of our countryside unions is reading this they might do the necessary sleuthing – I'm afraid I haven't time – to see whether the geocachers could be voluntarily confined to barracks each spring.

* * *

'Access' is a perennial challenge. We have a very old tradition of 'right to roam' in Scotland and there has never been a law of trespass despite the new Scottish Parliament pretending to bestow new rights in 2003. Then, a few years later, farmers were paid a subsidy to install footpaths. Taking the view that people were walking along our farm tracks anyway, and that it is better to channel walkers where they are not causing harm, I applied for the subsidy and put in some footpath signs. Then, surprise, surprise, the subsidy disappeared like snow off a dyke and we were left with the footpath. That didn't cause us any problems until we went back into dairying and that track was used by cattle again, so that the walkers had to pick their way through slurry and negotiate electric bungees. The council were soon on to me following complaints from irate ramblers. We had one of those conversations where the very nice Access Officer felt that he had to make me aware of public disapproval and I felt duty bound to make him aware that, whilst we had no problem with walkers, we were not going to change our farming practices to accommodate them. Furthermore, I told him our risk assessment said that the mixing of old ladies with yappy terriers and herds of cows in a confined space bounded by electric fences was asking for trouble. As it was now their 'core path' could they please absolve me in writing of any blame in the event of an accident? In the meantime I suggested a different circular route around the estate, which was safer and no less scenic. He said he would recommend to his boss that they should re-route the footpath to avoid council liability. The boss refused but the council have provided some helpful signs:

Caution
Please be aware these tracks are used daily by unattended
dairy cattle, which may be accompanied at certain times
of the year by bulls.
All fences are electrified to guide them to their grazing pastures.
Please leave all gates and bungees as you find them.
Please keep dogs under control and do not allow them to foul
pastures, please 'bag it and bin it'.
Let go of your dog if chased by cattle.

At least they have been warned.

* * *

The long-anticipated COP26 conference in Glasgow is upon us. The word 'crisis' has seemingly been attached to anything environmental during its protracted, Covid-postponed run-up, evoking anxiety and cynicism in equal measures. But for anyone with an interest in our rivers, we have clear evidence seen with our own eyes: salmon numbers are in sharp decline, despite numerous measures to help. Extinction of the wild Atlantic salmon during our lifetimes is a very real possibility. This is by any definition a crisis, not just for the species itself but because the salmon, as it migrates from small upland streams through river systems and estuaries to the Arctic Circle and back again, touches on more different species and environments than almost any other. As it grows from parr to smolt to grilse it feeds birds ranging from kingfishers to ospreys, and mammals such as otters, dolphins and seals. Humans also benefit from a healthy salmon population, and their long decline has almost made us forget the lost netsmen's jobs in coastal communities to add to the now precarious jobs on the rivers and in lodges, hotels and shops. Salmon need cold, clean water, food sources at critical stages and a balanced environment in which their predators are not so numerous that they inflict abnormal losses. And so focusing on the salmon as an indicator species helps us gain a holistic view of the whole ecosystem and help to restore it.

About the only thing that everyone previously agreed was that salmon numbers are falling to dangerously low levels and the reasons are multi-factorial. Every river is different, not just in its characteristics but also in the migratory routes and destination of its salmon. But we do know many of the causes, and mostly they can only be solved directly by governments and their agencies. We know that certain hydro-schemes, particularly large-scale ones installed in the mid-twentieth century,[71] interfere with upstream migration but also, far worse, kill many of the smolts during their fourteen-day downstream migration period. We know that open cage salmon farms spread disease and pollute coastal waters. We know that inappropriate forestry schemes acidify river water and we know that silty run-off from forestry and farming damages aquatic life. We also know that by bringing killer whales close to extinction we have allowed seal numbers to rise to unnatural levels. All of these problems can be solved.

The COP26 delegates can't fail to be made aware of the salmon's plight thanks to The Missing Salmon Alliance, which includes my friends Pedro Landale, Robbie Douglas-Miller and Tiggy Pettifer, and brings together passionate fishermen with the main UK Salmon conservation NGOs and charities, and has made great progress with scientific research by tracking salmon electronically (and could do more with extra funding). They have installed *Salmon School* – an artwork by Joe Rossano consisting of 350 glass salmon. It hangs in the delegates' dining room to act as a constant reminder of the fragile nature of the salmonid ecosystem. Let's hope they take action.

November 2021

It is a rewarding time for the dairy. When I was a beef farmer I used to dread it. We would watch as the pastures became poached, then, with heavy hearts, take the decision to bring the cows indoors for a winter of hard work and expense feeding and bedding them. We

---

71. Notably the Dee in Galloway and the Conon and its tributaries in Ross-shire.

converted to dairy and invested in cow tracks and electric fencing more in hope than expectation that the new mob grazing method would do what it promises – allow milk to be produced from grazing from February to December. Last week we had six inches of rain, bridges burst and Dumfries was flooded. We brought the cows in and fed them silage but within a couple of days they were out again. Fields that were underwater would previously have stayed submerged all winter, but no longer. A week later I drive across fields to feed pheasants without leaving a mark, held up by a dense mat of fibrous roots beneath the surface created by the continuous three-week cycle of grazing for a few hours by 650 cows in tightly packed paddocks. That is tangible carbon sequestration and something for which we should be rewarded not reprimanded.

The contrast with the arable block is stark. It was a race to lift the tatties and we just had the last trailer load going out of the gate as the heavens opened and turned the tattie field into a vision of Passchendaele. The geese have had a lovely time but any hopes of drilling the winter wheat we had planned are now fast receding. One of the principles of regenerative agriculture is that you should never leave soil uncovered by vegetation, and there is a risk that we will lose some topsoil as well as the nutrients in it. So it will prey on my conscience until there is a weather window to sow the wheat.

One consolation was that it allowed me to see how the land drains are working, and so for the first few days I was like a small boy playing Poohsticks, going around looking at where pipes emerge into ditches. One new ditch we had dug, to take the outflow from one of the new ponds, turned out to connect with a large drainage pipe that I never knew existed, so it was a big excitement to see water bubbling up into the ditch and a torrent heading towards the sea where I had never seen running water before.

* * *

As Covid restrictions are slowly lifted, anthropologists must be having a ball. The rituals of social intercourse, as various lost peoples emerge from the ulu to re-establish tribal bonds after a time of pestilence, are ripe for observation. My first major outing to London since the plague was for two memorial services at the

Guards' Chapel, neatly planned by the regimental adjutant to be either side of 'the Nulli' – our twice Covid-postponed regimental dinner. And so middle-aged men in blue-red-blue ties were to be seen squinting through the railings on Birdcage Walk to find fault with the Grenadiers mounting Queen's Guard, before sitting misty-eyed in the chapel as one of our tribal elders hauled himself into the pulpit with the aid of a treasured thumb stick to bid farewell to an old friend. 'There is now a gap in our ranks that can never be filled.' Then we ranged happily through clubland like a herd of elephants finding its favourite grove of marula trees again after a long absence. And the dinner was full of laughter and anecdote as we remembered good times and bad in war and peace, marvelled at how feral subalterns are now generals, captains of industry and men of the cloth, and made a valiant effort to drink two years' worth of claret. The Dire Straits ballad 'Brothers in Arms' is a great song:

> Someday, you'll return to
> Your valleys and your farms
> And you'll no longer burn
> To be brothers in arms.

but could not be more wrong.

Then, closer to home, we made a restored pilgrimage to the Church of St Vigean, better known these days as Ecclefechan, birthplace of Thomas Carlyle, and of the late lamented single malt, Fechan Whisky (try saying it in the local dialect after a couple of drams). The village hall is home to the renowned Fechan Hop where Dumfriesshire gentry gather for an annual feast of tribal dancing. There were some doubts whether herd muscle memory would be resilient enough to master the intricacies of Hamilton House and the Duke and Duchess of Edinburgh, but these were dispelled as soon as the band struck up and the tribe's familiarity with its cultural rites reasserted itself: 'Will you go first?' 'No we don't want to start.' 'Remind me what we do here.' 'Teapots, remember, ladies don't go down.' Then various social stereotypes reappeared; the chaps with militaristic leanings: 'Right, we are going to do Aberdonian, one, two, one, two ...' The testosterone-fuelled young lotharios vigorously tying their partners in extravagant knots. The man with a tin ear who claps a beat after everyone else. The giggling girl, fortified with a little too much Dutch courage, being gently prodded in the right direction as she sways down the set, trying hard to avoid a wardrobe malfunction. All watched with beaming contentment by dowagers gliding effortlessly across the floor like ships in full sail. And as we immersed ourselves in the music and surrendered to the familiar catechisms of the reels, and the steam gathered in the rafters, the tribe renewed its bonds of friendship through clasped hands and unspoken understandings and courtly gestures unchanged since Carlyle's day. Girls whom I have twirled and spun for the best part of fifty years set their jaws and prepared to give as good as they get with the same look of fierce determination they had when they were ten years old and at the tribe's ceremonial rite of passage, the All Ages Dance. And we all felt that normality had been restored, though a Martian, or even someone from Carlisle, would be astonished to hear it called that.

Epilogue

What a difference two years can make. In 2019 I was at the end of my tether with a failing business based on beef and holiday cottage tourism. Both sectors seemed doomed. Today, thanks to Covid – there had to be a silver lining somewhere – domestic tourism has never been better and the holiday cottages are full. Meanwhile Brexit has certainly been the disruption that we feared it might be, but in our case its lasting effects have so far been positive. It forced us to focus on the fundamentals and change our farming policy to what our farm is best suited to, rather than relying on subsidies. I won't make the same mistake again of writing that all will be well; farming has a habit of springing nasty surprises. But the dairy goes from strength to strength. Covid, followed by the invasion of Ukraine, has caused soaring inflation, which has for now brought an end to farm-gate food price deflation. Suddenly editors are commissioning articles on food security and meat consumption actually went up in Veganuary. John the bank manager, in his wisdom, made us budget on a milk price of twenty-five pence per litre, our latest budget has it at thirty-eight. And a return to pasture-based farming has allowed the wildlife on the estate to flourish.

My 'Fox-North coalition' with George Galloway was much longer than Fox and North's but was never destined to last. Putin's invasion of Ukraine caused me to sever the alliance. Typically, George had been trying to prevent war and showing that there are two sides to every story. He had long been warning that NATO should have allowed Russia to join the alliance and turn its face

East towards China and South towards militant Islamism. And he had warned that moves to bring Ukraine into NATO without Russia would provoke Putin. He may have been right about that but when Putin did invade I felt some of his comments were wrong, gratuitously contrarian and likely to be counter-productive, and I felt that I needed to distance myself from him and publicly broke our alliance. We parted on good terms.

The Alliance for Unity is now just a network of pro-British Scots. Next time there will have to be a better idea to defeat the SNP at the polls. Though Scotland is still held captive by the separatist cult, their star appears to be waning again. There are police investigations into allegations of SNP corruption. There is half-hearted talk of a further referendum but their plans for separation seem even less coherent than they did in 2014. And, in the event of Scexit, there is now wider support for the South of Scotland to break away with Orkney, Shetland and North East Scotland to form a separate devolved Scotland within the UK. We live in interesting times. But for now, here at Arbigland, there is milk and honey.

Acknowledgements

I would like to thank Andrew Johnston, now retired proprietor of Quiller Publishing for commissioning this book and for all the support he and his wife Gilly have given me over the years. And the new owner Nick Hayward and his team, especially Angeline Wilcox, Martyn Beardsley, Matthew Collis and Phillip Dean for putting the book together and marketing it. And the artist Guy Callaby for his wonderfully evocative cover. *Land of Milk and Honey* has grown out of my journalistic writing and I am very grateful to the following editors and their publications for commissioning me and for allowing me to reproduce text from some of my articles in the book: Patrick Galbraith, Editor of *Shooting Times*; Kate Green, Deputy Editor of *Country Life*; Tom Welsh, Comment Editor at the *Daily Telegraph* and the Comment Desk editors: Lucy Denyer, Maddy Grant, Harry Hodges, Sherelle Jacobs and Olivia Utley; Magnus Llewellin, Scottish Editor of *The Times*, Andrew Yates, Features Editor of the *Daily Mail*; Will Moore, Features Editor at *The Spectator*, Graham Stewart, Deputy Editor of *The Critic*; Michael Mosbacher, former Editor at *Standpoint* and *The Critic*; Mark and Mary Devlin, co-owners and editors at *TheMajority.Scot*; and Brian Monteith, Editor of *Think Scotland*. The parts of the book about 'wilding' would not have been possible without the research paid for by the Charles Douglas-Home Memorial Trust and I am very grateful to Jessica Douglas-Home and the other trustees for making me a generous award. It is a wonderful institution that supports journalism, long may it flourish. I am also very grateful to Charlie Burrell, Isabella Tree, James Rebanks, Derek Gow, Jim Lowther, Tom Leicester, Jake Fiennes, Innes Macneill, Lynn Cassells and Sandra Baer for their time. The parts of the book about the Alliance for Unity would not have happened without George Galloway, James Giles and Jean-Anne Mitchell and all our wonderful candidates and I thank them for their support.

And last, but most of all, my family Sheri, Oliver and Rosie and the great team we have here at Arbigland, particularly Davie McWilliam, Graham Kennedy, John Rippon and my new partners in the dairy, Richard Beattie, Brendan Muldowney and Joey Malone.

Thank you.

Jamie Blackett
Arbigland
March 2022